I0034448

AN INTEGRATED APPROACH TO HOME SECURITY AND SAFETY SYSTEMS

AN INTEGRATED APPROACH TO HOME SECURITY AND SAFETY SYSTEMS

Sonali Goyal
Neera Batra
N.K. Batra

CRC Press
Taylor & Francis Group
Boca Raton London New York

CRC Press is an imprint of the
Taylor & Francis Group, an **informa** business

MATLAB® is a trademark of The MathWorks, Inc. and is used with permission. The MathWorks does not warrant the accuracy of the text or exercises in this book. This book's use or discussion of MATLAB® software or related products does not constitute endorsement or sponsorship by The MathWorks of a particular pedagogical approach or particular use of the MATLAB® software.

First Edition published 2022
by CRC Press
6000 Broken Sound Parkway NW, Suite 300, Boca Raton, FL 33487-2742

and by CRC Press
2 Park Square, Milton Park, Abingdon, Oxon, OX14 4RN

© 2022 Taylor & Francis Group, LLC

CRC Press is an imprint of Taylor & Francis Group, LLC

The right of Sonali Goyal and Neera Batra to be identified as author of this work has been asserted by him in accordance with sections 77 and 78 of the Copyright, Designs and Patents Act 1988.

Reasonable efforts have been made to publish reliable data and information, but the author and publisher cannot assume responsibility for the validity of all materials or the consequences of their use. The authors and publishers have attempted to trace the copyright holders of all material reproduced in this publication and apologize to copyright holders if permission to publish in this form has not been obtained. If any copyright material has not been acknowledged please write and let us know so we may rectify in any future reprint.

Except as permitted under U.S. Copyright Law, no part of this book may be reprinted, reproduced, transmitted, or utilized in any form by any electronic, mechanical, or other means, now known or hereafter invented, including photocopying, microfilming and recording, or in any information storage or retrieval system, without written permission from the publishers.

For permission to photocopy or use material electronically from this work, access www.copyright.com or contact the Copyright Clearance Center, Inc. (CCC), 222 Rosewood Drive, Danvers, MA 01923, 978-750-8400. For works that are not available on CCC please contact mpkbookspermissions@tandf.co.uk

Trademark notice: Product or corporate names may be trademarks or registered trademarks and are used only for identification and explanation without intent to infringe.

ISBN: 978-0-367-63841-2 (hbk)
ISBN: 978-1-032-11138-4 (pbk)
ISBN: 978-1-003-12093-3 (ebk)

DOI: 10.1201/9781003120933

Typeset in Sabon
by MPS Limited, Dehradun

Contents

Preface

An efficient and accurate home security and access control to the doors security system, which is based on face and voice recognition, is very important for a wide range of security applications. Security is an important aspect or feature in smart home applications. Most countries are gradually adopting smart door security systems. The most important major part of any door security system is accurately identifying the individuals who enter through the door. Face with voice recognition is probably the most natural way to perform authentication between human beings. Additionally, these are the most popular biometric authentication traits, after fingerprint technology. Biometric recognition, or simply biometrics, is the science of establishing the identity of a person based on physical or behavioral attributes. It is a rapidly evolving field with applications ranging from securely accessing one's computer to gaining entry into a country. While the deployment of large-scale biometric systems in both commercial and government applications has increased the public awareness of this technology, *An Integrated Approach to Home Security and Safety Systems* is the textbook to introduce the fundamentals of biometrics with integrated algorithms of face and voice recognition in order to provide security to an individual who is living alone at home. The two commonly used modalities in the biometrics field, namely face and voice, are covered in detail in this book. A few other modalities, such as hand geometry, ear and gait, are also discussed briefly along with advanced topics, such as fingertip application for heartbeat detection.

Designed for undergraduate and graduate students in computer science and electrical engineering, *"An Integrated Approach to Home Security and Safety Systems"* is also suitable for researchers and biometric and computer security professionals.

MATLAB® is a registered trademark of The MathWorks, Inc. For product information, please contact:
The MathWorks, Inc.
3 Apple Hill Drive
Natick, MA 01760-2098 USA
Tel: 508-647-7000
Fax: 508-647-7001
E-mail: info@mathworks.com
Web: www.mathworks.com

Acknowledgments

First and foremost, praises and thanks to God, the Almighty, for His showers of blessings throughout our journey to complete this book successfully. You have given us the power to believe in our passion and pursue our dreams. We could never have done this without the faith we have in you, the Almighty.

We are extremely grateful to our parents for their love, prayers, caring and sacrifices for educating and preparing us for our future. Finally, our thanks go to all the people who have supported us to complete the book work directly or indirectly.

List of abbreviations

AI	Artificial Intelligence
ANN	Artificial Neural Network
ARM	Advanced RISC Machines
ATM	Automated Teller Machine
BCG	Bacille Calmette Guerin
BP	Blood Pressure
BPM	Beats Per Minute
BSN	Body Sensor Network
CDMA	Code Division Multiple Access
CMOS	Complementary Metal Oxide Semiconductor
DCT	Discrete Cosine Transform
DFT	Discrete Fourier Transform
DNA	Deoxy Ribonucleic Acid
DRAM	Dynamic Random Access Memory
DSS	Decision Support System
DWT	Discrete Wavelet Transform
ECG	Electrocardiogram
EISPACK	Eigen System Package
FFNN	Feed Forward Neural Network
FFT	Fast Fourier Transform
FIR	Finite Impulse Response
FOG	Freezing of Gait
GA	Genetic Algorithm
GPS	Global Positioning System
GSM	Global Source of Mobile Communication
GUI	Graphical User Interface
HMM	Hidden Markov Model
ICA	Independent Component Analysis
IIR	Infinite Impulse Response
IoT	Internet of Things
IP	Internet Protocol
IT	Information Technology

IVR	Interactive Voice Response
JPG	Joint Photographic Group
LDA	Linear Discriminant Analysis
LID	Language Identification
LINPACK	Linear System package
LLRO	Local Linear Regression
LPC	Linear Predictive Coding
MAC	Media Access Control
MEMS	Micro Electro Mechanical Systems
MFCC	Mel Frequency Cepstral Coefficient
MFRASTA	Mel Frequency Relative Sptectral
MICS	Medical Implant Communication Service
MLP	Multi Layer Perceptron
MRA	Multi Resolution Analysis
NN	Neural Network
PAUSN	Privacy Aware Ubiquitous Social Networking
PC	Personal Computer
PCA	Principle Component Analysis
PDA	Personal Digital Assistant
PGM	Portable Gray Map
PICA	Principle Independent Component Analysis
PIN	Personal Identification Number
PLDA	Programmable Linear Discriminant Analysis
PLP	Perceptual Linear Prediction
PTT	Pulse Transit Time
RASTA-PLP	Relative Spectral Perceptual Linear Prediction
RBF	Radial Basis Function
RFID	Radio Frequency Identification
RGB	Red Green Blue
SMS	Short Message Service
SOM	Self Organizing Map
SRAM	Static Random Access Memory
SVM	Support Vector Machine
UI	User Interface
VTLN	Vocal Tract Length Normalization
Wi-Fi	Wireless Fidelity
ZCPA	Zero-Crossing Peak Amplitude

Chapter 1

Overview of pervasive computing

1.1 OVERVIEW OF PERVASIVE COMPUTING

Pervasive computing is the ability to use information part and applications of software anytime, at any place by any person. If comparison is done between this computing and desktop-type computing, pervasive computing may access data by preferring any type of device, in any type of location with any format [1]. Desktop computing suffers mainly with two problems:

Limited accessibility: In this, the network of an organization can't be accessed by all the users of an organization. Maximum number of users can't afford a desktop computer, and it is also not possible for organizations to provide a computer to every member.

Isolation access: Isolation property means desktop computer is meant for single user use in a particular location. Users try to interact with a number of groups. So pervasive computing comes into a role in the way to address these challenges.

Definition of pervasive computing: This computing was defined by Mark Weiser developed at research and development in different applications such as mobile computing, context-aware computing, and wireless communication. It is the profound technology that always disappears [2]. It is basically a technology that works with the collection of small devices integrated into real-life objects and provides an omnipresent access for the computations. The aim of ubiquitous computing is to layout infrastructure of computing in a way that they combine in a seamless manner with the surroundings and become invisible. Mark Weiser defines three categories for computations: pads, boards and tabs. As the size of tabs was very small for writing notes, screen was very sensitive and awareness of location was also an issue. So, the pads came into the role. Their functionality was similar to tabs but size was larger in case of pads and it was a notebook. Last is boards, size of the board was larger as compared to tabs and pads. Basic property used with pads, tabs and boards includes mobile devices, context awareness, wireless communication, etc. They incorporate themselves into everyday's life fabric until the devices are equivalent to them. It is also

DOI: 10.1201/9781003120933-1

termed as omnipresent computing. The aim of this computing is to be everywhere at all the time. To be everywhere, this computing will have to support number of devices with changing hardware capabilities. For more coverage area, this computing must be compatible with number of types of networks. Networks include the GSM (Global System for Mobile communication), CDMA (Code Division Multiple Access) network for mobile phone services, wireless network technology, Bluetooth, etc. These devices support connections for different distance ranges. Bluetooth network provides connection in 10–100 m. Wireless technology supports large distances with the help of access points. Mobile phone network supports distance up to 35 km. A property called switching between networks will need session management. The overview of pervasive computing technique is given in Figure 1.1.

Pervasive computing can exist in several forms, stemming from the use of laptops to household appliances. Some of the technologies that make it possible are microprocessors, mobile codes, sensors and the Internet. In short, pervasive computing happens every time when people use digital devices to connect to technological platforms. The main goal of pervasive computing is to embed computation into an environment that allows users to enjoy every day objects' benefits through information processing. Pervasive systems are expected to perform in a broad set of environments

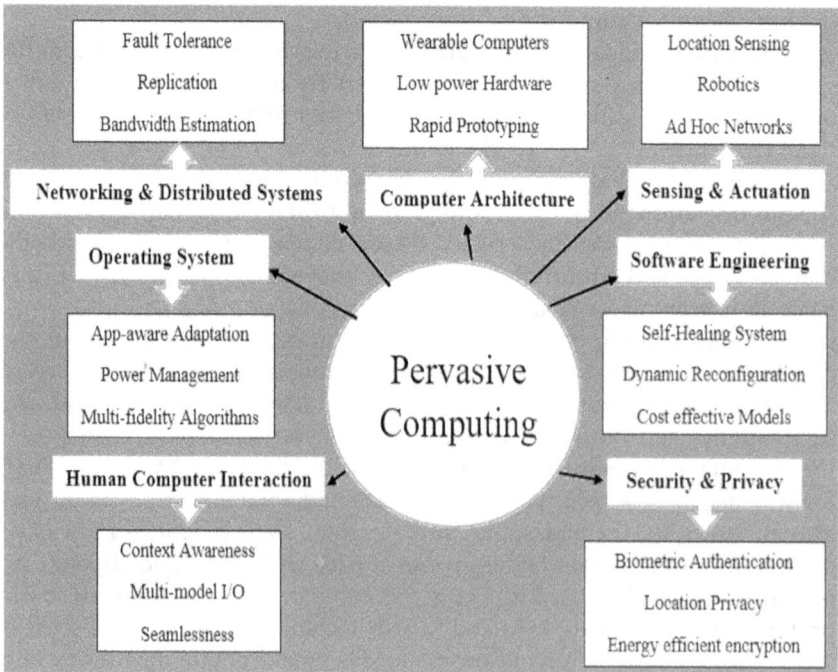

Figure 1.1 Overview of pervasive computing.

with different capabilities and resources. Application requirements may change dynamically requiring flexible adaptation. Sensing faults appear during their lifetime and as users are not expected to have technical skills, the system needs to be self-managing.

1.2 DISTRIBUTED COMPUTING

A distributed system is a collection of independent components located on different machines that share messages with each other in order to achieve common goals. From 1970s to 1990s, a research has been followed to create a framework that will work with interconnection of two computers by a network. Network can be static or dynamic, wired or wireless or it can be pervasive. It is a system used to coordinate the processes that are performed upon number of computers simultaneously [3]. Medium of communication between the computers is through the messages as shown in Figure 1.2. This computing is used to solve complex problems that can't be solved on a single computer within a particular time. Software is required to coordinate the task in distributed computing, which is a big challenge because main task is to divide the elements that can run individually. In distributed computing, special software is required to manage the resources of a number of computers. This software is run by a main computer unit called as 'master computer', and other computers are known as 'slaves'. Master computer is responsible to allocate the tasks to slave computers and to combine the results when processing is complete.

One drawback of using a master -slave system is the possibility of failure at a single point, i.e., server. As it is the only computer that provides necessary resources to all the slave computers, if the server goes down, the system will

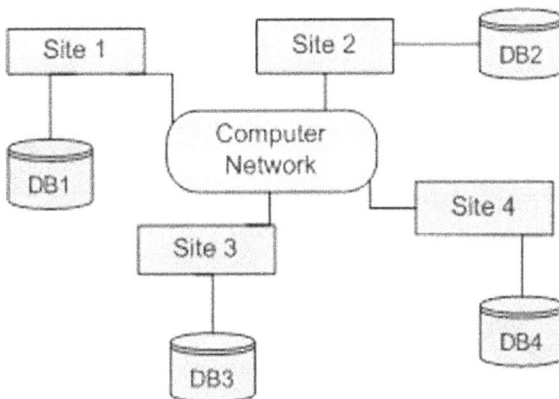

Figure 1.2 Overview of distributed computing.

not work properly, and another drawback is that if the number of slave computers are extremely high, in that case, the resources will be scarce.

1.3 MOBILE COMPUTING

In the 1990s, when issues emerged in building a distributed system, around then, another exploration has been finished with wireless local area networks and laptop computers, i.e., mobile computing came into existence. Various standards or elements of principles or functions of distributed computing are applied ceaselessly in day-to-day life. Yet, a few imperatives are there because of which new technology is developed, constraints are trusted with low value, network quality variation, robustness of elements, size problems, battery constraints and weight limitations [4]. Mobile computing is a type of computing that is used while we are in a moving state. It is a technology that allows transmission of data, voice and video via a computer or any other wireless- enabled device without having to be connected to a fixed physical link as shown in Figure 1.3. It requires wireless communication to facilitate mobility of devices in communication mode.

This technology has a number of benefits: Productivity is increased by using mobile devices in a number of companies and thereby cost and time of the clients is reduced. Another benefit is that it can be used for entertainment purpose (video and audio recordings can be streamed on-the-go using mobile computing), portability (a user can work without being in a fixed position and work from anywhere as long as there is a connection established) and cloud computing (it is the ability to access data anytime, anywhere if Internet is connected and then files can be accessed on number of mobile devices) as shown in Figure 1.4. Cloud computing services cover a vast range of options now, from the basics of storage, networking and

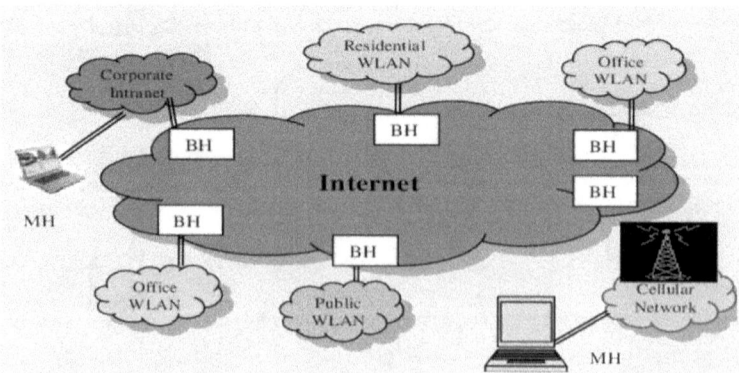

Figure 1.3 Overview of mobile computing.

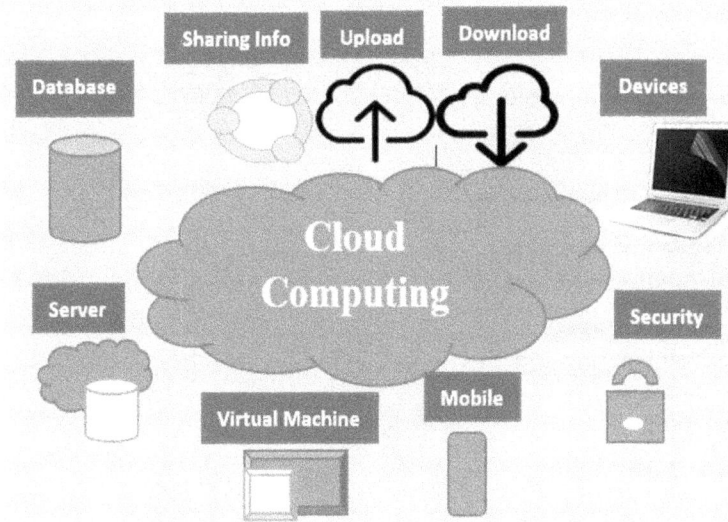

Figure 1.4 Overview of cloud computing.

processing power through to natural language processing and artificial intelligence as well as standard office applications.

1.4 PERVASIVE COMPUTING

Pervasive computing is the area that is growing very rapidly in today's world. In this technology, small devices are allowed to be used in omnipresent way and embedded in an invisible manner in order to provide a ubiquitous environment. By Weiser, a system called as pervasive computing creates a fully connected medium that is integrated with normal environment and that is not distinguishable from that. Pervasive computing devices are very small devices in a range of a few millimeters to small meters, and these devices are connected to each other via wired or wireless links [5]. The aim of this computing is to make life simpler for every individual by using a number of tools, so that, information can be managed successfully and easily. Pervasive computing technique is beyond the concept of personal computers. It is based on the concept that from clothing to any type of devices, they are integrated with chips in a way to connect with the network of other devices. These devices consist of a number of characteristics like: they have small, inexpensive processors with limited memory, these devices are connecting with other devices without any user interaction and they will be connected by wired or wireless links. Schematic series of pervasive computing is depicted in Figure 1.5.

```
┌──────────────────────────────┐
│ Remote Communication         │
│ Fault Tolerance              │      Distributed ┌──┐ Mobile     ┌──┐ Pervasive
│ High Availability            │      Systems     │  │ Computing  │  │ Computing
│ Remote Information Access     │                  └──┘            └──┘
│ Distributed Security          │
└──────────────────────────────┘

┌──────────────────────────────┐
│ Mobile Networking            │
│ Mobile Information Access     │
│ Adaptive Applications        │
│ Location Sensitivity         │
│ Energy Aware Systems         │
└──────────────────────────────┘

┌──────────────────────────────┐
│ Smart Spaces                 │
│ Invisibility                 │
│ Localised Scalability         │
│ Uneven Conditioning          │
└──────────────────────────────┘
```

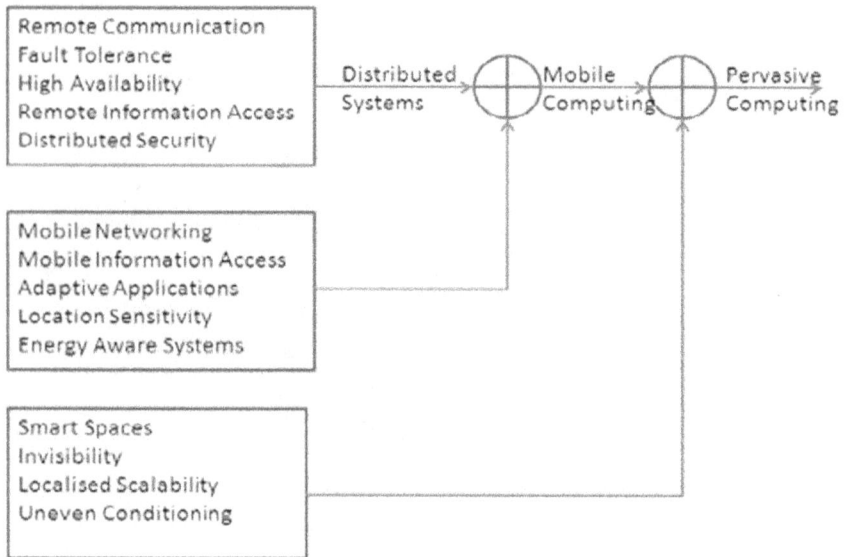

Figure 1.5 Schematic series of pervasive computing.

1.4.1 Pervasive computing examples

Pervasive computing comes with number of names like 'Ubiquitous Computing', 'Augmented Reality', 'Virtual Reality', 'Wearable Computing', 'Ambient Computing', 'Things That Think' and 'Smart Space Computing'. In routine life, if a person realizes that a box is too heavy to lift, person whistles or calls for a super-heavy helper to assist that person. But in pervasive computing, if a person goes to lift the block and invisible computer agent detects that person is not strong enough to do so, it automatically assists that person without even asking for it.

In the next scenario, there is a room that automatically manages the heating, cooling, and lighting effects based on a user's profile. The other is 'smart' or 'intelligent refrigerator' as shown in Figure 1.6, which is used to warn the owner when eatables are expired or out of date. This refrigerator should be aware of the contents within the refrigerator, and it must have the ability to order new food when supplies are low, but this smart refrigerator is not available in the market yet.

The technology that is smart and used by everyone is 'smart phone'. A smart phone is a mobile phone that includes advanced functionality beyond making phone calls and sending text messages. Most smart phones have the capability to display photos, play videos, check and send e-mail and surf the Web. Today's mobile devices are packed with nearly 14 sensors that produce raw data on motion, location and the environment around us as shown in Figure 1.7.

In mobile computing, a person having cell phone with call facility and message facility remains connected while he/she is moving from one area to

Figure 1.6 Smart refrigerator.

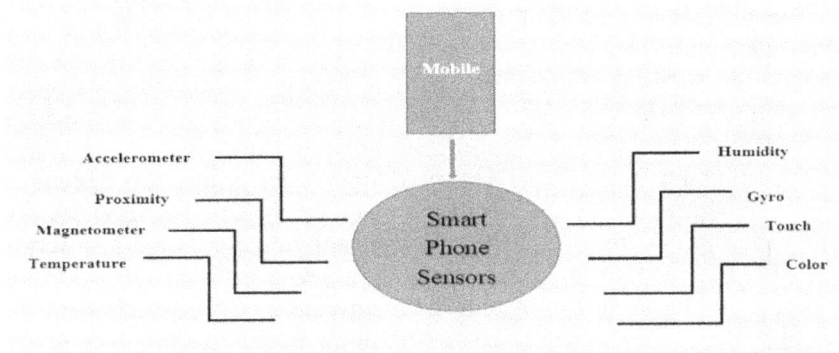

Figure 1.7 Sensors used in a smart phone.

another. But in pervasive computing, a person having smart phone remains connected and informed while passing the refrigerator even, when he/she is ready to leave the home, that, butter is almost finished. So, when he/she is passing through a grocery store, message is received on his/her glasses that displays 'Buy Butter' message. Areas where this computing can be used are mobile devices, wireless communications, motor traffic, context-aware computing, military, production, smart home, E-commerce, medical technology, etc., as shown in Figure 1.8.

1.4.2 Pervasive healthcare system

In traditional healthcare system, critical patients mostly don't get treatment in time that results in high death rate and the patients who are treated by

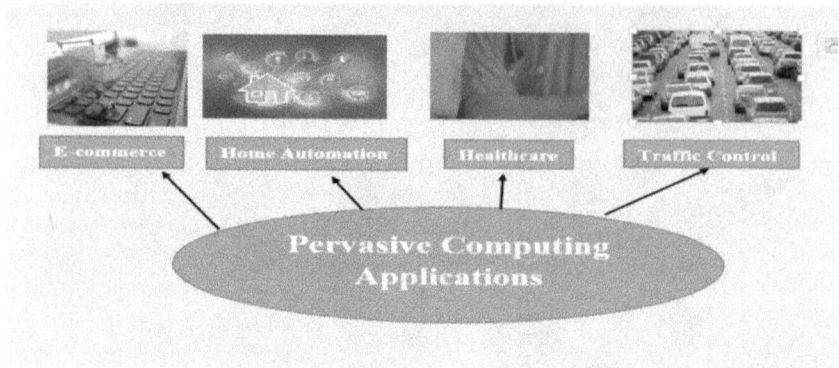

Figure 1.8 Pervasive computing applications.

one doctor are not entertained by other doctors due to privatization of hospitals [6]. Many times the ambulance carrying patients gets stuck in the traffic resulting in delayed treatment and hence death of the patients. Moreover, on the way to hospital, only limited resources are available in the ambulance for the immediate treatment of the patient which can worsen the health condition of the critical patient. In pervasive healthcare system, these all issues are well dealt with, and security and safety of the patient is the major concern that is taken into consideration utmost.

The complete scenario of real-world pervasive healthcare system is shown in Figure 1.9. It examines the vision of new field pervasive healthcare and identifies new research thrusts: convenient communication and access to PHR (Patient Health Record), helping experts, patients and others to navigate and locate nearest health care, convenient traffic management for health experts. This system consists of multiple levels, each of which serves a different yet unique purpose. Each module works in an independent way, yet they are interconnected by the centralized network and server. The system is divided into three parts: centralized network and server, the smart ambulance and the user application/interface. Each of them consists of various modules, functioning in an autonomous way and serving a different purpose, all interconnected. The smart ambulance has a traffic monitoring module, which has been implemented using smart IoT (Internet of Things) technologies. What this module does is, it takes the real-time location data and gives a shortest distance from ambulance to the patient and further to the nearest hospital. The smart ambulance has some embedded health sensors in it, which help to monitor the patient's condition on the go. There are various sensors that will be giving live reading of the patient's health to be sent directly to the concerned doctor. The medical team connected to the centralized network and the general physician who stays next to the patient in the ambulance, they keep on monitoring the patient's health condition continuously in real time through sensors embedded in the ambulance and

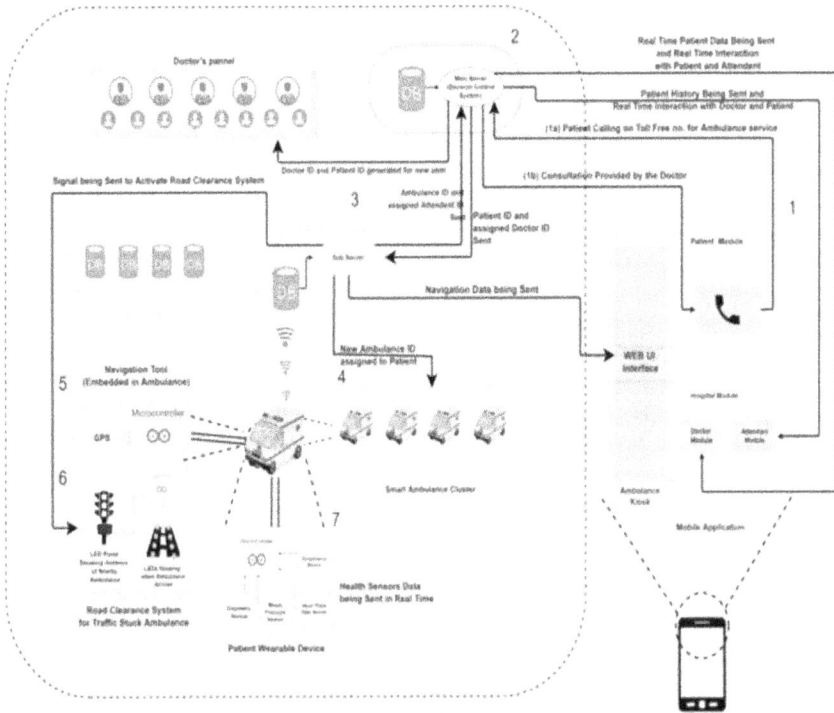

Figure 1.9 Pervasive computing healthcare system-example.

wearables worn by patient until he/she reaches safely to the nearest hospital. This helps the associated doctors, present at the hospital, to make necessary arrangements before the patient arrives the hospital. Also, the general physician and the medical team take care of patient, if his/her health worsens in the journey before arrival to the hospital.

1.4.3 Challenges of pervasive computing

1. **Context awareness:** One should always be context aware so that they can know about user's state and surroundings and also have the ability to modify their behavior [7]. A user's context encompasses both human factors as well as physical environments. Human factors have factors such as heart state, temperature and emotional state, while physical environment has factors such as noise and location. Context free is well defined by five W's. The 'five W's' include the following:
   - **Who:** the ability of devices to identify not only their owner but also other people and devices in its vicinity within the environment.

- **What**: the ability to interpret user activity and behavior using that information to infer what the user wants to do and provide the necessary information and help.
- **Where**: the ability to interpret the location of the user and use that to tailor functionality. This one is the most explored aspects of context.
- **When**: the ability to understand the passage of time, use it to understand the activities around and to make inferences.
- **Why**: the ability to understand the reasons behind certain user actions. This might involve sensing the user's affective state, such as body temperature and heart rate.

2. **Social impact:** Pervasive computing will have a more critical social impact contrasted with regular processing innovation inside any climate, changing our associations with gadgets as well as others also. The implanting of computational antiquities and increase of basic ancient rarities with no computational abilities except for exist inside the climate, will bring about better approaches for acting and communicating among individuals in that climate. This will affect individuals' current social conduct, most likely bringing about new social structures. The ramifications of this will be that we are not just keen on relational connections among co-found individuals (without the innovation) or connections among individuals and gadgets just (within the sight of the innovation), however in the framework intervened human–human connections that arise within the sight of the innovation.

3. **Privacy and trust:** Privacy is one of the most social issues for users and security of information. It results in continuous monitoring of user's activities. The main factor to depend on pervasive computing infrastructure is to trust it. Conversely, infrastructure needs to be confident of user's identity.

    As client turns out to be more reliant on a pervasive figuring framework, thusly, it turns out to be more educated about that client's inclinations, personal conduct standards, developments and propensities. Basic models incorporate user identity, which may be shown up through the validation of his own cell phone, the area of places he has been in a day and so forth. This brings up issues of who can approach such data and what they could utilize it for. Clients who have a decent comprehension of such frameworks and like the conceivable potential for the deficiency of protection may pick not to utilize them, since in corrupt hands; this data may be transformed into a weapon against them. To get clients to depend on pervasive figuring framework, we should get them to confide in it. On the other hand, the foundation should be sure of a client's personality and approval level to react properly to their solicitations. The test at that point, is to build up this common trust in a way that is negligibly meddling and consequently, safeguards imperceptibility. Examination issues, here, include: different strategies that could be utilized for client verification in this worldview, deciding if we could repeat techniques that were created for the work area climate or rather centred on more up to date and apparently more precise biometric techniques. We should adjust the necessity for consistent conduct and the need to caution clients about like.

4. **Device heterogeneity and resource constraints:** Ubiquitous computing will have to support multiple devices with varying hardware capabilities in order to be everywhere. These capabilities result in trade-offs that inevitably influence the development of applications and their capabilities. With multiple devices at each user's disposal, there will be a need for mechanisms that help determine which tasks are appropriate for each device. As users move from one device to the next, applications with the ability to move from one device to the next will need to readjust to the uniqueness of each device, optimizing their effectiveness according to the device capabilities. As the devices get more versatile, more imperatives emerge. They have to be more modest in nature to allow users to conveniently move around with them; this size constraint limits other resources like screen size, processing power, battery life and so forth, which in turn influences other factors like connectivity and the development of services and applications.

5. **Power management:** Power consumption consists of three parts: storage, communication and processing. To control power, a technique named 'turning a knob' is used. It is easy to lower the power of wireless transmitter by reducing output power, effectively reducing its range. Another method is switching between heterogeneous subsystems, e.g., switching between Wi-Fi and Bluetooth networks. But support to deal with heterogeneous interfaces, more software support is required. Processing is a part whose requirements varied from little to extremely high. If the issue is of single processor, it is possible to control power consumption by either selectively deactivating individual block. Power consumption can also be controlled by using multiple processors. Storage is the system that can consume power in pervasive system. There are number of storage media that are available for such systems: physical disk, DRAM, SRAM integrated with processor. Similar to processor and wireless subsystem, each kind of storage presents a different power profile to the system.

6. **Wireless discovery:** Management is also a big issue for the given system itself. In the previous time, there were number of people using single computer. Then, with the PC revolution, a stage is reached when single computer is used by a single person, now the time has come with the introduction of cell phones and other portable devices; a stage is reached now where multiple computing devices are

associated with a single person. As per person system usage increased, the conceptual, physical and virtual management of these devices becomes an issue. Embedded processing in everyday objects such as a coffee cup or chair exasperates this problem: number of computing devices that are growing exponentially, they must somehow be managed. Major problem that is faced for large collection of devices is just a basic knowledge of what exists: if I know something exists, how do I find it or if I have a collection of devices, how do I know what they are? The solution to this problem is to release objects from a specific designation and treat a coffee cup as just a coffee cup instead of a specific coffee cup. This shift makes our life easier. Computer systems are having a particular unique name, i.e., IP address, which is a convenience that may not exist in a deeply embedded environment.

7. **User interface adaptation:** A feature of pervasive computing systems is that they merge a collection of devices with very tiny sensors to palm, notebook and computers, each with different size. Applications require working effectively in a heterogeneous environment by the system's organizer point of view and the users must have full control of each device or component with a limitation of their sizes [8]. For example, tiny devices have tiny displays and requires the user to navigate a series of these menus. The problem is with PDAs; they are heavily loaded with features but never used because of their complexities.

8. **Location aware computing:** The feature of ubiquitous computing when compared to conventional computing is the use of location. When Internet is considered, a server is not aware of the client location but when computing embedded is used into the local environment, interaction can be advanced to improve the result. For example, a query about the nearby equipment can be automatically fulfilled by the current location. Also the information accessed at a given location on one occasion can be offered to other people, if same query exists. In this, if an example of conference room is taken, then links are provided on the whiteboard automatically that have been used by previous groups [9]. In the same way, low-priority messages are not allowed to be delivered to a laptop, if a meeting is going on side by side. In universities, context systems are designed by the tool kits.

1.4.4 Issues in pervasive computing

Various issues, their effects in respective areas and other possibilities in corresponding areas have been taken into account and represented here in the form as shown in Table 1.1.

Table 1.1 Issues, their effects and possibilities in different areas

Issues	Effects	Other possibilities
Suddenly ill, lost consciousness, fell down from sofa	a. Detects by impact sensor b. Inform the physician.	High BP, heartbeat, temperature
If not in mobile state from a time period higher than a threshold value	a. Detects by wearable sensors b. Alarm should be raised by sensors	Orientation and movement of person may be examined
Presence of body fluid around a person	a. Check chance of bleeding	Urine or vomit fluid may be there
Detects a rest or fix posture of a person which remains unchanged over a time frame	a. Detects by posture sensor	
Abnormal heart rate	a. Detect by wearable sensor b. Inform the physician	Abnormal B.P
Amount of medicine has been decreased	a. Use counter value b. Order for medicine at medical store	When 2 days left for medicine, message should sent to doctor and patient
If he has not left the house for a number of days	a. Detects by motion sensor b. Alerts are sent out	Almira has been open or not for a number of days
If resident has been in bed for more than 12 hours	a. Detects by pressure sensor in bed mattress b. Alerts are sent out	
Room temperature should be within a specified range (12–30) if outside this range for more than a specified period	a. Detects by temperature sensor b. Residents will be alerted	Concept of windows (if open permanently at low temperature, alerts should be sent)

1.4.5 Advantages of pervasive computing

- Complexity of new technologies is removed.
- Information can be accessed conveniently.
- Information is managed easily, efficiently and effortlessly.
- Invisibility (smart environment will be integrated with network technologies that are invisible to the user).
- Decision making (better choices can be made about everyday things by using smart environment).

- Convergence (interconnected network technology is included in the environment so due to sharing number of problems can be eliminated).

1.4.6 Disadvantages of pervasive computing

- Network connection is slow in case of this computing.
- Operating cost is highly expensive.
- It is not a secure method so trust-based models are required for higher security.

1.5 BIOMETRICS

Latest advancements in the era of IT on security and authentication move forward to biometric system. These two factors: authentication and security are required in real-life applications, such as ATM transactions, law enforcement, banking, military, passport immigration, and offices/buildings, where everyone has to authenticate his/her identity [10]. Previously, the methods used for user authentication were login ID/password, smart card, PIN identification, etc. But they had some drawbacks associated with them like PIN thefts, stolen card, forgot password, etc. So to overcome these problems of traditional methods, biometric technology came into light.

Biometrics are physical or social human attributes that can be utilized to carefully distinguish an individual to give admittance to frameworks, gadgets or information. Instances of these biometric identifiers are fingerprints, facial examples, voice or composing rhythm. Every one of these identifiers is viewed as one of a kind to the individual, and they might be utilized in blend to guarantee more prominent exactness of ID. Since biometrics can give a sensible degree of trust in verifying an individual with less grating for the client, it can possibly significantly improve endeavor security. PCs and gadgets can open naturally when they recognize the fingerprints of an endorsed client. Worker room entryways can swing open when they perceive the essences of confided in framework executives. Help work area frameworks may consequently pull up all pertinent data when they perceive a representative's voice on the help line.

Biometrics term is composed of two words: Bio+metrics that mean life measure. There are a number of systems that require authentication in order to determine the individual's identity. Main role of these schemes is to allow data to a valid user not fake one. It is used to identify the identity of a person by using some methods, and there are number of methods available for this technology: a) face recognition, b) voice recognition, c) iris recognition, d) palm recognition, e) fingerprint recognition and f) DNA. Today, DNA examines are utilized basically in law requirement to distinguish suspects – and

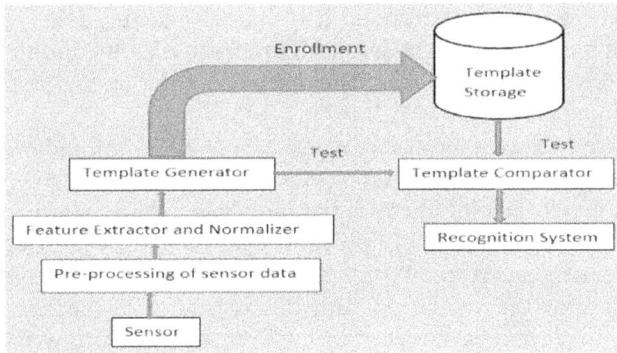

Figure 1.10 Biometrics architecture.

in the motion pictures. Practically speaking, DNA sequencing has been excessively delayed for far reaching use. This is beginning to change. A year ago, a scanner hit the market that can do a DNA coordinate in minutes. Biometric systems are automated systems for identifying the user through physiological or behavioral characteristics. Physiological method is if the shape of the body is considered for identifying a person. From more than 100 years, the oldest features are fingerprints that have been used for recognition purpose. Other examples are face recognition, ear recognition and iris recognition. Behavioral method is related to the behavior of a person for the recognition purpose (Figure 1.10).

1.5.1 Biometrics architecture

Biometric system architecture has the following main components [11]:

1. Sensor
2. Preprocessing of sensor data
3. Feature extractor and normalizer
4. Template generator
5. Template comparator
6. Recognition system

The above biometric framework segments and their working are clarified as given beneath:

1. **Sensor**: Sensor is the main square of the biometric framework that gathers all the significant information for the biometrics. It is the interface between the framework and the present reality. Ordinarily, it is a picture-securing framework; yet, it relies upon the highlights or attributes necessitated that it needs to supplant or not.

2. **Preprocessing of sensor data:** It is the second square that executes all the prepreparing. Its capacity is to improve the info and to dispose of the curios from the sensor, foundation commotion and so on. It plays out some sort of standardization.

3. **Feature extractor and normalizer:** This is the third and the main advance in the biometric framework. Extraction of highlights is to be done to recognizing them at the later stage. The objective of an element extractor is to describe an item to be perceived by estimations.

4. **Template generator:** Layout generator produces the formats that are utilized for the validation with the assistance of the removed highlights. A layout is a vector of numbers or a picture with particular plots. Attributes got from the source bunches together to frame a layout. Layouts are being put away in the information base for examination and fills in as a contribution for matcher.

5. **Template comparator:** The coordinating stage is being performed by the utilization of a matcher. In this part, the acquired format is given to a matcher that contrasts it and the put away layouts utilizing different calculations, for example, hamming distance and so on. Subsequent to coordinating of the data sources, the outcomes will be produced.

6. **Recognition system:** It is the gadget that utilizes the consequences of the biometric framework. Iris acknowledgment framework and facial acknowledgment framework are some basic instances of utilization gadgets.

1.5.2 Commonly used biometrics methods

There are number of biometric methods in use as shown in Figure 1.11. Biometric has been for long the target of future authentication that expected that biometric authentication will largely displace other means of our current authentication and access control. Biometric authentication techniques are classified by the type of characteristics evaluated: physiological attributes or behavioral singularities.

- **Physiological biometrics:** These are based on classifying a person according to data obtained as part of the human body such as his/her fingerprints, face or eye iris.

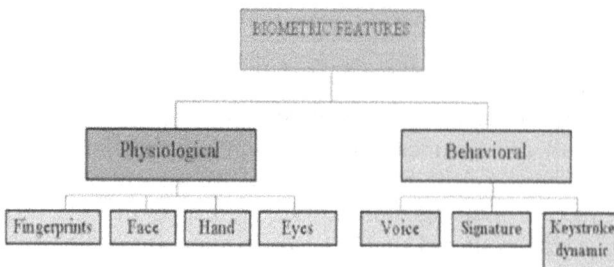

Figure 1.11 Classification of biometric features.

1.5.2.1 Fingerprint

It is one of the biometric technologies used in number of applications. It is used to identify a user uniquely using patterns found on fingertip [12]. A fingerprint consists of a pattern of valleys followed by ridges placed on each person's fingertip.

An uneven surface with ridges and valleys on fingertip makes a pattern unique as depicted in Figure 1.12a and b. Fingerprints were used for authentication of a user and the matching accuracy was very high. Previously, method used was old ink and paper based but now an automatic version is generated for identification purpose. In this, the user has to place his finger for the print to be electronically read. If user operates in a controlled environment, then this technique is best. Arrangement of ridge makes unique fingerprints for every user, and it requires less memory for data storage but in case of dryness, skin infection and dirty skin sometimes it makes wrong identification and it is a noisy sensor.

1.5.2.2 Face recognition system

Face recognition includes several landmark points that make facial features. User's face can be captured by a camera for authentication purpose. It is an application for automatically recognizing a user from camera [13]. Its main advantage is that it can be used to recognize a user from a particular distance. This system is secure and easily implemented but variation in pose and lighting affects the image. As there are number of features included with face recognition technique, it can be used in various areas such as banks, criminal's identification and identification of terrorists.

It is a biometric approach used to distinguish one person from other on the basis of their faces. It involves three main steps as shown in Figure 1.13.

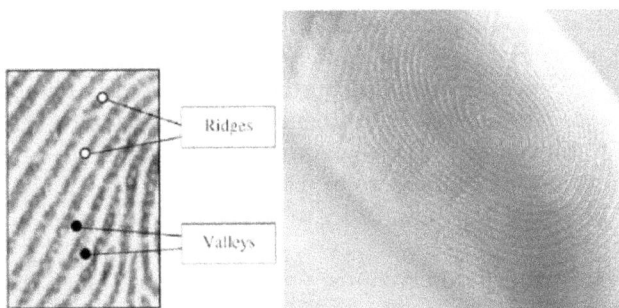

Figure 1.12 a) Image of ridges and valleys in a finger, b) real image of ridges and valleys in a finger.

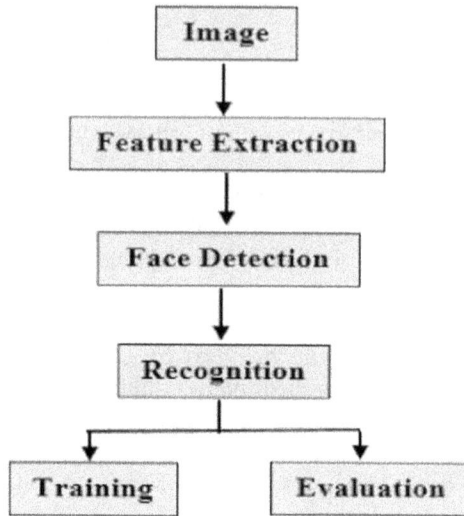

Figure 1.13 Face recognition steps.

- **Data acquisition:** Acquisition of data items is the main step for recognition of face in which facial images are collected by using number of sources. Number of images when collected will be having different pose variation, illumination variation, etc. So we can easily verify the performance of our recognition system with the number of conditions. Sometimes, input image is normalized and different transformation methods for image can be applied on input image. In this, unwanted parameters such as noise, lighting and blurred effects are removed and then it will be used for next process.
- **Feature extraction:** Feature extraction is the next step of this system, which is used to take out the information by using given images and then extracted features are used for recognition step. In this step, dimensionality reduction is used with noise cleaning.
- **Recognition:** Recognition is the next step of face recognition system after features selection and extraction, now image is classified by using number of classifiers. In this step, number of images is taken and feature extraction is applied on them and stored in database. When new image comes for face recognition, it first performs face acquisition, feature extraction then compare all the images in database with this new image. An image to be tested is compared with all the images stored in the database in order to identify an individual.

Automatic face recognition comprises mainly of two tasks:

- **Verification:** In verification part, the system correlates the given person with who that individual says they are and gives the output in the form of yes or no decision.

- **Identification:** In identification part, the system matches the given person with all the other persons stored in a database and provides a list of matched persons.

1.5.2.2.1 Challenges of face recognition

There exist a number of challenges correlated with recognition of facial features and they can be defined by the following entities:

- **Pose:** Variation in pose is a matter of concern in recognition of face, pose changes nearby camera when the person is at the door, at that time, some features may visible partially or fully closed as shown in Figure 1.14 and it results in wrong identification [13].
- **Structural components:** Some features are there like moustaches, scarf, beards and glasses, which may or may not be always present (Figure 1.15), and this problem will affect the classification process.
- **Image rotation:** Face images fluctuate or differ by various rotations.
- **Poor quality:** Certain images that may contain noise or are blurred or distorted may result in a poor quality of face recognition.
- **Facial expression:** Face expressions may change from person to person as shown in Figure 1.16.
- **Unnatural intensity:** Faces from three-dimensional models such as rendered faces, animated movies and cartoon faces have unnatural intensity.

Figure 1.14 Variation in pose.

Figure 1.15 Variation in structural components.

1.5.2.2.2 Face recognition algorithms

Face is life's most important visual object that can be used for conveying identity. Humans basically use faces to detect individuals. The methods used for face recognition are holistic, structural and hybrid methods. In holistic approach, the complete face region is taken into account as input data into face matching system. One of the best examples of holistic methods are Eigenfaces (most widely used method for face recognition), principal component analysis (PCA), linear discriminant analysis (LCA) and independent component analysis (ICA), etc. It uses global representations or we can say descriptions for full face irrespective of some local features of face. This method is further divided into two parts: Statistical and AI. In statistical, images are represented as two-dimensional arrays and the main concept used is to calculate correlation parameter between input image and other stored database images. This method is very expensive as it does not work on variable conditions like pose, lighting, etc. and second drawback is it performs classification always in high dimension space. To overcome these problems, other methods can be used in order to gain more accurate and meaningful results.

- **Principle Component Analysis (PCA) Algorithm**

It is very popular algorithm developed in 1991 presented by Turk and Pentland. PCA algorithm depends on Eigenvector concept [14]. By using this algorithm, images are taken for the analysis, and these images must be of same size. First, a normalization technique is applied to mark the features of eyes and mouth from a face. PCA algorithm is used to compress the image that's why it is said that it depends on dimensionality reduction concept. From reduction, the information that is not required is removed and then decomposes the structure of face into number of uncorrelated components which are known as Eigenfaces. Every stored image will be combined by a linear sum of weighted Eigen feature vectors which are known as Eigen faces. At the time of recognition, new image is projected onto subspace and Euclidean method is used for calculating distance for

Figure 1.16 Facial expression variation.

classification purpose [15]. The aim of using PCA is to collect important features of data by using a matrix method [16]. In AI, quantities of instruments are utilized to perceive face like neural organization and machine procedures. It implies when guideline segments are removed from PCA calculation, at that point, neural organization is utilized as a classifier to lessen parts into low measurement space. SVM is a calculation utilized for issues of example grouping. Gee is additionally utilized for this. Its disadvantage is it doesn't work under high inconstancy in posture, light condition and so on.

Advantages:

- It is simplest, easy and efficient way of face recognition.
- It works well in less or more pose or illumination variation.
- It uses the concept of Euclidean distance, and it is based on the dimensionality reduction process, which reduces the dimensions of original recognition while retaining the most of information.
- It reduces the noise factor as the maximum variation basis is chosen and so the small variation in background ignored automatically.
- It is based on multiple images taken as input concept.
- It takes less computational time.

Disadvantages:

- It is expensive for high-dimensional data.
- It does not implement in proper way when there is large variation in pose, orientation, illumination, etc.

- **Independent Component Analysis (ICA)**

This algorithm is used to analyze large order data or we can say multivariable data. The main technology that is used in this algorithm is conversion of high-dimensional space into low-dimensional space in such a way that new space which is containing converted variables will define the essential features of data [17]. Cluster analysis is a technique that is used in this algorithm. A group of number of data elements is considered as a cluster and cluster analysis is defined as a technique used for allocating space in a region to data where large concentration is present and that region is called as cluster.

- **Linear Discriminant Analysis (LDA)**

LDA is a technique used for separation of data items. It is an approach used for classification of samples in the case of unknown classes that is based on samples of training using classes of known users. The aim of using LDA is to maximize variance across the users and minimize variance within the users. It basically makes clusters of the images of classes that are similar and partitioned the images of classes which are dissimilar to each other. This technique is used only for classification, but it can't be used for regression. In order to recognize an input image, input image is compared to each and every image that has been stored in the database then closest distance is calculated and on the basis of distance input image is identified. It can be used in pattern recognition, statistics and machine learning. It is also known as Fisher-face- based algorithm that depends on dimensionality reduction concept.

Advantages:

- It maximizes the class distance but distance is minimized within the class.
- It is robust.

Disadvantages:

- It does not consider multiple images as input.

- **Discrete Cosine Transform (DCT)**

It is used to express a sequence of points for data items working at different frequencies in terms of sum of cosine functions. This transform is used mainly for compression standard in which system receives face as an input image, then normalized or crop that face then to calculate distance between faces and finally comparison is done.

Advantages:

- It works well when there is less variation.

Disadvantages:

- It works well under same orientation, pose, illumination, etc.
- It is not having high success rate.

- **Gabor Wavelet**

It characterizes the image as localized orientation selective and frequency selective features [18]. This technique is basically a filter used for face detection and it is named after Dennis Gabor [19].

Advantages:

- It works well in large variation in pose and illumination conditions.

1.5.2.3 Hand geometry

This method is very popular these days as it works well in most situations for user recognition. This method automatically measures the number of dimensions of the hand and fingers then compares those measurements to a template [12]. When a person places his hand on the surface embedded with sensors and uses poles in between the fingers for proper placement of the hand, a reading is started, and this procedure comes under a term called as spatial geometry. Finger or hand geometry generally measures two or three fingers as shown in Figure 1.17. This method has an advantage that it does not require any co-operation from the user for identification but problem is its size which is very large.

1.5.2.4 Iris scan

This method is used to identify user by using different patterns of human iris. Iris is a muscle inside the eye that is used to control the light which is entered into the eye; it will regulate the size of pupil as shown in Figure 1.18. It is an accurate method among other biometric systems as it gives high reliability and high recognition rate. It does not require any close contact between user and machine. Irises of identical twins are different which an advantage of this system is.

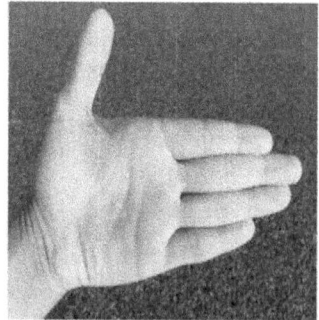

Figure 1.17 Recognition based on hand geometry.

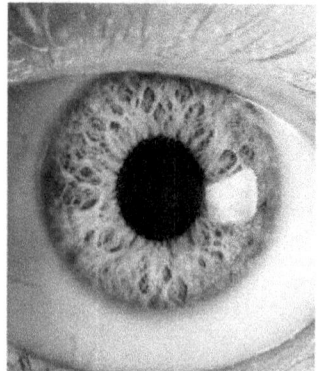

Figure 1.18 Recognition based on iris.

- **Behavioural biometrics:** It consists of measurements taken from the user's actions, some of them indirectly measured from the human body.

1.5.2.5 Keystroke dynamics

It is an automated method used with the help of keyboard for examining the keystrokes of a person. By this method, dynamics of a person's keystrokes are examined like speed and pressure, when a user pressed or released a key. The aim of this technology is to improve robustness and uniqueness. An example of using this application is the use of computer access, in which this biometric technology can be used to recognize the computer person's identity continuously. Persons are there to authenticate themselves on system by entering their login id and password. The advantage of using this technique is it does not require any additional device except keyboard which is globally available.

1.5.2.6 Dynamic signature verification

This method is one of the most demandable form of user's verification in number of applications such as document authentication and financial transactions. This method is used to examine the speed, direction, accuracy and pressure of writing as shown in Figure 1.19. In offline mode or previously used method, users use to write their signature on normal paper, then digitize that paper using a scanner and finally, system use to verify the signature but in online mode as used now a days, users save their signature in iPad or digitizing tablet by using PDA's and then recognition is done.

This system is not expensive to use, and it requires very less time to verify the user but problem with this technique is that its error rate is high.

1.5.2.7 Speech/voice recognition

This technique is used to recognize a user on the basis of voice signals as it transforms voice into text. For voice recognition, a number of strategies can be used: support vector machine, Hidden Markov Model, dynamic time

Figure 1.19 Recognition based on signature.

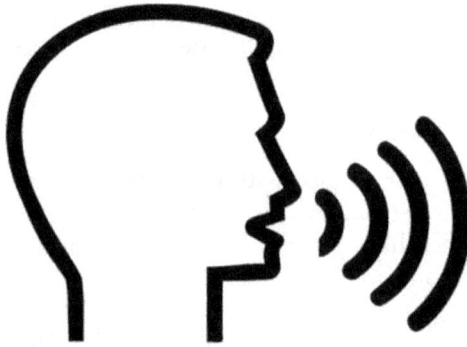

Figure 1.20 Voice signal generated by a word spoken by a person.

warping, vector quantization, etc. By using microphone, audio files are recorded [20]. But noise factor is the major concern in this biometric technique. It takes less time for user authentication, but its accuracy is low. Figure 1.20 shows a signal generated by a word spoken by a person.

1.5.2.7.1 Voice recognition system

Voice is a name given to sound that carries the content of language. Voice recognition is a task of verifying a person on the basis of waves included in the voice. Voice of a person is identified in order to verify the identity and used in the number of applications such as voice dialing, voice mail and database access services. When a person speaks into microphone, at that time, computer converts the sound of words into another form, i.e., text that is available on computer system. The computer or laptop either repeats the words spoken by the person or it will provide a prompt for next turn. In earlier times, voice recognition programs use to make a person to speak in staccato fashion; this fashion insists a person to leave a gap between two words but now voice recognition systems are very easy to use. A person can speak without any pause between the words. The only condition is person must speak in a clear way. In this system, two sessions are included:

- Enrolment session
- Operation session

In enrolment session, the users who are already registered will provide voice samples in a way to train the model for that user.

In operation session, matching is done by taking voice sample of all the users stored in a database. Finally, decision is taken in order to accept or reject the identity of that user.

All the steps of voice recognition process are shown in Figure 1.21.

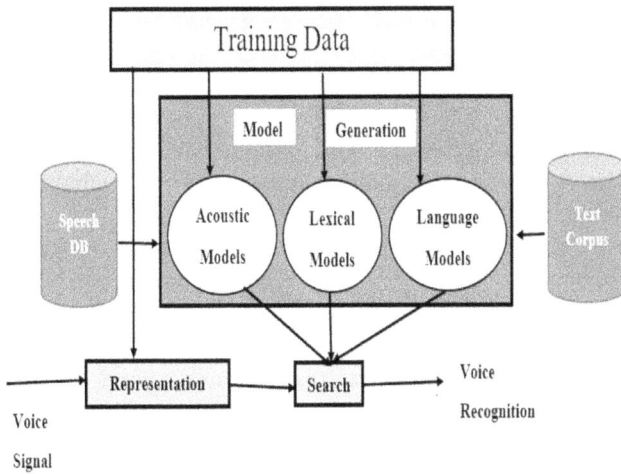

Figure 1.21 Voice recognition process.

- **Representation**
  Preprocessing: It is the first step in recognition system in which voice is recorded by using a microphone, then signal is captured at a sampling frequency of 5000 Hz. In this step, frequencies that are below 100 Hz are removed by using high-pass filters (these frequencies were containing noise factor). Important factor performed in preprocessing is segmentation or end point detection. Speech segmentation starts when the energy of signals exceeds the threshold value and ends when the energy of signals drops below the threshold value.
- **Model**
  Feature extraction: After processing of signals, next step is extraction or taking out the features, and for calculation purpose, observations are taken out with time frames. It is to be assumed that speech signal will be constant within these frames with a length of 25 m. The time frames are overlapping and shifted by typically 10 m. On the time window, a fast Fourier transformation is performed, moving into the spectral domain. Human ears do not perceive all frequency bands equally. This effect can be simulated with bandpass filters of non-uniform frequency band widths. Until 500 Hz, the width of the filters is 100 Hz, after that it increases logarithmically. The filter centre frequencies are defined in the so-called Mel scale. The spectrum is decorrelated with a discrete cosine transformation. Of the resulting coefficients, the first coefficients carry the most significance. Therefore, only the first, e.g., ten coefficients are selected as feature vector. The resulting features are called Mel Cepstra, commonly abbreviated as MFCC.

- Search
  i. **Decoding:** It is a way to calculate the matching rate of sequence of words with signals generated by the feature vectors. In the process of decoding, there are three sources of information which are available: a) acoustic model, b) dictionary and c) language model. The condition for decoding is to know the probability of words that can be spoken.
  ii. **Postprocessing:** It is the last step of speech recognition process in which signals that are processed already are now classified in acceptance or rejection decision.

1.5.2.7.2 Voice recognition techniques

- **Mel Frequency Cepstral Coefficient (MFCC)**

It is the most commonly used cepstral analysis technology used for recognition purpose. In this, logarithmic representation is used for the frequency bands so it provides clearer response than other systems. In this, feature vectors of MFCC are computed by each frame present in the analysis of signals [21]. First, coefficients are extracted by the signals taken as an input; these signals are divided into frames. On these frames, when hamming window is applied, discontinuities of the signals are minimized at that time as shown in Figure 1.22.

Advantages:

- It is used for speech processing tasks.
- Its recognition accuracy and performance rate is high.
- It has low complexity.

Disadvantages:

- It includes little noise robustness.

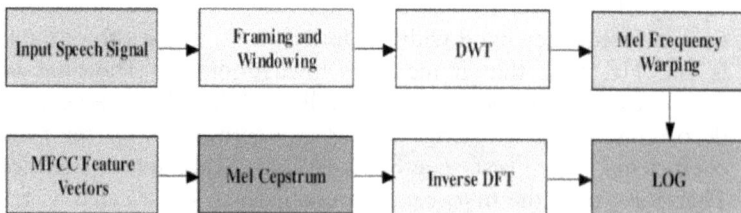

Figure 1.22 Steps of MFCC technique.

• Perceptual Linear Prediction (PLP)

It works in a similar way as linear prediction coding; it means it is also based on short-term spectrum of speech as shown in Figure 1.23. It eliminates unwanted information of signals of speech and helps in improving the recognition rate of speech signals.

• Linear Prediction Code (LPC)

It provides an accurate estimate of signals for speech processing. Speech is modeled in an efficient way by using this method [24]. It is assumed in this technique that all the speech signals are combined by the combination of past signals of speech. In this, initially, signals are segmented into number of frames then each frame will be multiplied by a window and finally the output will be passed through an auto correlation parameter as shown in Figure 1.24.

Advantages:

• It is a static technique [22].
• It is reliable, accurate and robust technique for providing parameters of speech.

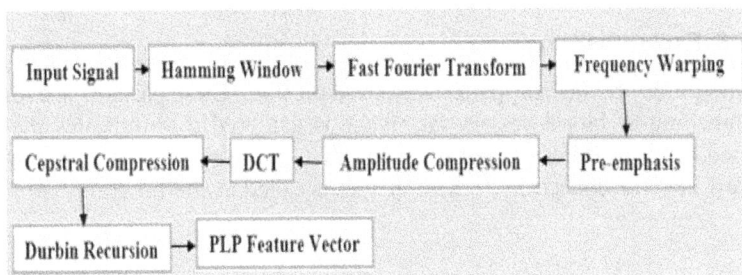

Figure 1.23 Steps of PLP technique.

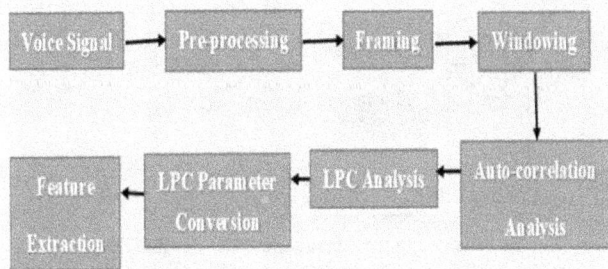

Figure 1.24 Steps of LPC technique.

Disadvantages:

- It is not able to distinguish the words with similar vowel sound.
- It generates residual error as output, resulting in poor speech quality.
- It includes little noise robustness [23].

- **RelAtive SpecTrA- Perceptual Linear Prediction (RASTA-PLP)**

This method is used to enhance the speech when recorded in a noisy environment. When combined with PLP, it gives better performance. It was proposed by Hynek Hermansky, it is a technique which is mainly used when signals are embedded with noise factor and when slow as well as fast varying information is to be removed from the speech signals [25]. It works in a robust way means there is no restriction on the distance taken for hardware devices to record the audio files as shown in Figure 1.25.

Advantages:

- It is robust.

Disadvantages:

- It gives poor performance in clean speech environment.

1.5.2.8 Facial thermograms

Thermograms are the heat patterns emitted on skin. These patterns are formed by branching of blood vessels. An image is gathered that indicates the heat radiated from the different parts of the body using infrared sensor as shown in Figure 1.26. The collection of these images is called as thermograms. Any part

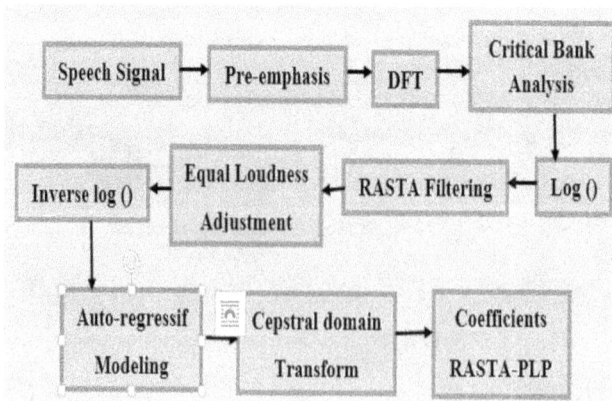

Figure 1.25 Steps of RASTA-PLP technique.

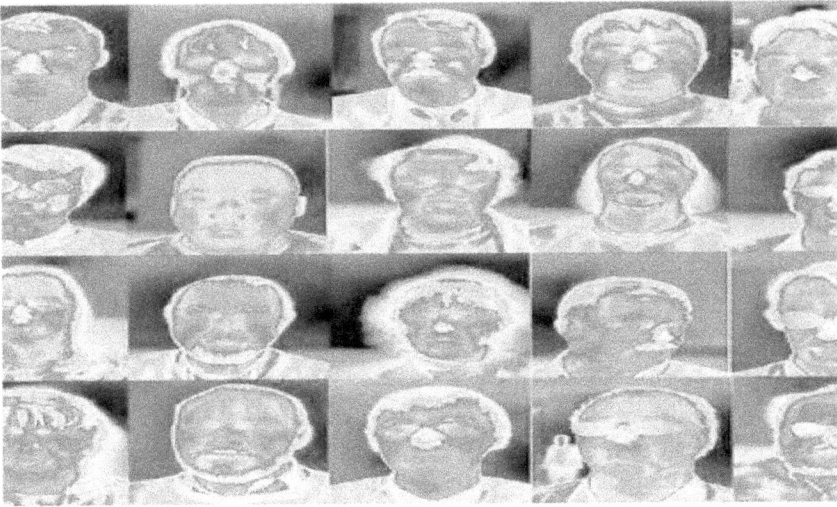

Figure 1.26 Detection based on facial thermograms.

of the body could be used for identification. This technique uses infrared and efficiently works in darkness, but it is affected by ambient temperature.

1.5.2.9 Ear

Human ear is used in biometric technology for authentication. Feature vector extraction is the main step of this recognition. It is known that the distinctive features are the shape of the ear and the structure of the ear as shown in Figure 1.27. But it is expected that features of an ear are not unique for every individual. Feature vector expression is based on geometry, internal structure of ear and on algebra such as PCA, ICA, etc.

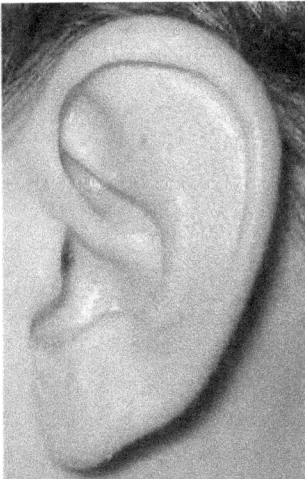

Figure 1.27 Ear-based recognition.

1.5.2.10 DNA

DNA (deoxyribonucleic acid) is used for user authentication. By this method, a unique code is provided for every individual as depicts in Figure 1.28. The fact is identical DNA pattern exists for the identical twins. It provides secure transmission of large data with failure rate 0%. Its benefit is that it is error free and its accuracy is high but problem with this technique is it is highly expensive.

Face and voice recognition techniques are the most reliable and accurate techniques used for user's recognition. This system is based on acquisition of data by an individual, extracting features of that data and then finally compares the extracted features with actual features stored in a database. These systems are used to increase user's security and convenience. For example, on airport, a user enrolled by inserting card at gate and see into camera then that camera will acquire his image of eye, process that image and compute the code for that image. Final step is to compare both the codes, one is actual code and other is computed code in order to verify the user's identity. It means every biometric technique has four parameters:

- Sensor or acquisition of data
- Feature extraction
- Matching
- Decision control system

In data acquisition, data is collected of an individual from number of sources. In feature extraction, acquired information is processed in order to extract the features. In matching parameter, extracted features are compared with features stored in the database and in last step, decision is taken in order to accept or reject the identity of an individual.

If we use different recognition methods then their summary is available in Table 1.2.

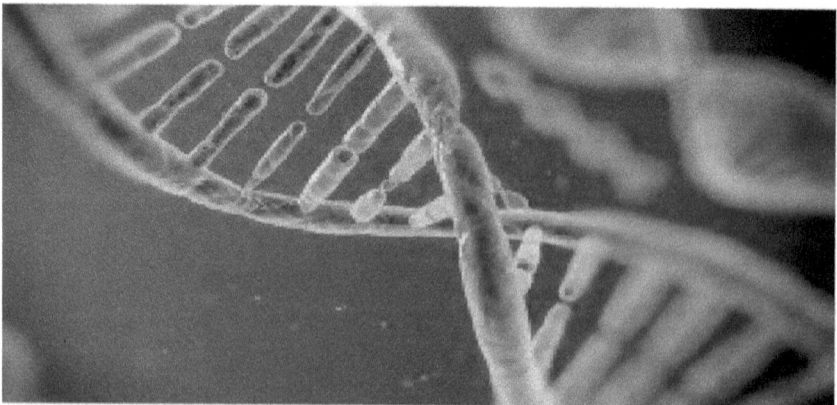

Figure 1.28 Structure of DNA.

Table 1.2 Different biometric methods followed with accuracy rate

Technology	Accuracy	Application
Fingerprint	High	Law enforcement
Face	High	Passport
Iris	High	ATM
Signature	Medium	Financial
Voice	Medium	Access control
Keystroke	Low	Computer security

Automatic biometric system is used now due to advancement in computer area. Automatic verification of any individual is based on the behavior of that person with some other characteristics.

1.5.3 Advantages

- **Universality:** The main feature of biometric technique is to be universal. Most people have their fingers, they can speak and their cells contain the DNA. There exists some exceptions. Some people who don't have their finger or arm, they cannot use fingerprint-based systems, etc. However, all of us have the DNA. The problem is inability of some people. But this problem can be solved by a multi-biometric system.
- **Uniqueness:** The biometric systems are very reliable as they are unique with the authentication process of the security system. The features are extracted from number of parts of the human body like the DNA, a fingerprint, etc. But the uniqueness is the most important part. It is not possible in every case but it holds mostly true.
- **Low circumvention:** The biometric security systems are very safe and secure. Most of the biometric features are possible to take by the authorized person so that his presence with the authorization device placed is must. In some cases, a person could also misguide the security system, i.e., a person can cut off a finger and tries to be accepted by the system. But now days a scanner is able to measure blood flow in the veins in the scanned finger. So, the system can't make fool by the person. This is how it is very safe and secure.
- **Scalability:** The biometrical technologies can be built as a multilevel authentication system. This multilevel system can include a standard login, a voice login, a fingerprint login and many others. Combining number of technologies in single unit increases the security of the complete system. If any one level of the whole system is broken, the others could decrease a break-through possibility.
- **Permanence:** It is a set of features acquired from the DNA as it is permanent and does not change. This same thing could be said of the fingerprints. But in case of the fingerprints, a man could cut his finger,

which would change the fingerprint structure. Next considered permanent biometric feature is a retina scan.

1.5.4 Disadvantages

- **Exactingness:** Some of the biometric features can be very exacting and unique to acquire, e.g., the DNA. Though the DNA is unique for all of us (apart from the twins), but it is very difficult to extract and to analyze it frequently. The price of the DNA analysis is not low. This excludes the DNA from the real-time applications; it is not worth the advantages it can bring these days.
- **Difficult implementation:** It is not an easy task to implement a reliable biometric system. The teams of researcher's work regularly for a safe and reliable implementation of the biometric system. Some of the parts of the system have already been done as finger print recognition, face recognition and specially the case of voice recognition. The security till date is not high.
- **Cooperation unwillingness:** Some persons are not willing with acquiring their biometrical features. Most of the people dislike scanning their retinas, scanning their fingerprints and faces and the least of them dislike recording of their voices. It is clear then that it would be very useful to develop a reliable technology based upon the voice processing.
- **Inconstancy:** Some of the biometric features are permanent but some are not as well. As the human voice is not fixed during the whole life. This can be the greatest problem. Some other features may also have instability like height.

1.5.5 Biometrics applications

- **Justice and law enforcement:** Associations like the Federal Bureau of Investigations (FBI) and interpol have been utilizing biometrics in criminal examinations for quite a long time. Today, biometrics is broadly utilized by law requirement organizations over the world for the ID of lawbreakers. Biometrics is likewise broadly utilized for jail and prison management. Biometrics gives a cutting edge arrangement by which the Jail Authority, Public Safety Departments and Governments can securely and safely oversee detainee characters.
- **Airport security:** Biometrics streamlines the air terminal experience for a great many travelers voyaging each day. Making the excursion through air terminal terminals more consistent for travelers is an objective mutual via air terminals around the globe. Biometric innovation to check traveler characters has been utilized in a few

enormous worldwide air terminals for various years and the innovation is rapidly spreading to different areas over the globe.

In numerous air terminals, the top biometric methodology decision for movement control is iris acknowledgment. To utilize iris acknowledgment, voyagers are first selected by having a photograph of their iris and face caught by a camera. At that point, their extraordinary subtleties are put away in a global information base for quick, precise ID at ports of passage and leave that utilization iris acknowledgment for voyager personality confirmation. When voyaging, rather than holding up in long lines to be prepared, travelers just stroll into a corner and investigate an iris camera. The camera at that point photos the iris and a product program at that point coordinates the subtleties with the data put away on the information base.

Biometrics streamlines the air terminal experience for a great many travelers voyaging each day.

• **Workforce management:** It is another field where the utilization of biometrics is on the ascent. Deceitful employee time and attendance exercises are a typical wonder in associations all through the world. As per an American Payroll Association study, the normal representative allegedly takes around 4 and a half hours out of each week, which is comparable to about a month and a half's get-away whenever extrapolated longer than a year. To unravel this issue, organizations are executing biometric time tickers on their work locales.

A biometric time and participation framework are the mechanized technique for perceiving a worker dependent on a physiological or social trademark. The most widely recognized biometric highlights utilized for worker ID are faces, fingerprints, finger veins, palm veins, irises and voice designs. At the point when a representative endeavors recognizable proof by their natural qualities, a biometric equipment gadget thinks about the new sweep to all accessible formats to locate a careful match. Indeed, even government associations presently depend on biometrics for guaranteeing convenient participation of staff and exact finance calculations.

• **Banking:** Banks are using biometrics to develop next-generation identification controls that combat fraud, make transactions more secure and enhance the customer experience. For example, in 2019, The Royal Bank of Scotland (RBS) declared a pilot of instalment cards including biometric unique mark innovation. The preliminary is being done with around 200 of the bank's NatWest clients and will happen in the United Kingdom. The unique mark goes about as a substitution for PIN passage and is utilized to confirm exchanges in abundance of £30, making it faster and simpler for clients to finish their instalments. This most recent advancement follows a long queue of biometrics development, as RBS and NatWest were the principal U.K. banks to empower Touch ID unique mark acknowledgment on their portable banking applications. ATM's are also more secured by using biometric technology. At present, numbers of other applications are there like internet banking, telephone banking that are totally secure with this technology.

- **Physical systems:** There are number of areas where biometric technique can be applied in order to provide maximum security in schools, military areas, etc. It is an essential part for parents, employees, etc. Fingerprint recognition technology in the biometric market has held the largest market size worldwide and has been widely adopted by many industries including schools. Fingerprint recognition is the most pervasive, old, simple, and cheap form of biometric technology. Although palm vein recognition, iris recognition and face recognition have been implemented in schools, finger scanning is by far the most commonly used technology in the U.S. education market. In the United Kimgdom primarily the type of biometric employed is a fingerprint scan or thumbprint scan but vein and iris scanning systems are also in use.

  In the military, the main enlightening protection concerns identify with the danger of capacity creep and the following abilities of biometrics. Capacity creep, or mission creep, is the cycle by which the first reason for getting the data is extended to incorporate purposes other than the one initially expressed. Capacity creep can happen with or without the information or arrangement of the individual giving the information. Numerous security specialists fight that capacity creep is unavoidable. Following, which might be idea of as a specific sort of capacity creep, alludes to the capacity to screen progressively a person's activities or to look through data sets that contain data about these activities. For instance, if an individual must utilize a standard biometric for various administrative, business and recreation exchanges of regular daily existence, it gets conceivable that every one of these records could be connected through the normalized biometric. This connection could permit an element, for example, the public authority, to arrange a thorough profile of the person's activities.

- **Immigration:** In whole world, illegal immigration is considered as a major threat in security. These are the automatic processes used for law breaks. To avoid such scenarios, the Government of Canada has started using biometrics. The government says that the collection of biometrics will facilitate application processing and simplify entry into Canada for low-risk travellers. Fingerprints and a photograph will get required for some individuals from Europe, the Middle East and Africa applying for a Canadian guest visa, work or study license, perpetual living arrangement or shelter in Canada. Biometrics are used at both the application and entry into Canada phases. Biometrics allow visa officers to screen applicants for prior criminal convictions or Canadian immigration infractio0ns. A traveller's biometrics are also used when they enter Canada to confirm his or her identity.

- **Healthcare:** Biometrics technologies, such as fingerprint scanners, palm vein readers, facial recognition tech, iris scanners and others, have long held promise to tighten up identification of patients and employees. This would help dependably confirm that patients are who they state they are, ensure guardians are working with the correct clinical and segment data and guarantee just the best possible representatives approach the correct data. Be that as it may, biometrics innovations have been delayed to get on in medical services. Without a doubt, numerous emergency clinics and centers have executed some fundamental tech; however, biometrics is not yet completely in the standard of medical care rehearses. What's more, it will even now require some investment and exertion with respect to medical care CISOs and biometrics innovation merchants to get the tech swimming in the standard.

1.6 SENSOR-BASED TECHNOLOGY

This technology consists of a sensing chip used to sense vital parameters from a person's body. It converts one form of energy into another form and converts physical parameter into electrically measureable signal [26]. A sensor is used to convert the physical parameters such as temperature, heartbeat, sweat rate and blood pressure into another form of signals that are measured electrically. Example of temperature is taken to explain this technology. When liquid is expanded and contracted by the use of mercury in glass thermometer, at that time reading is measured by the person. There are a number of sensors used electromagnetic, thermal, optical, etc. In previous time, the size of sensors was very large so it can't be used in the form of wearable technology. But nowadays, microcontroller circuits, amplifiers and functions make wearable sensors to be used for health monitoring. Sensors and wearables are attached into number of accessories such as cups, garments, belt, wrist band, eyeglasses and shoes. Wearable sensors are found to be useful because of advantage such as use of sensors for data collection which when attached to human body continuously monitor patient's health thereby reducing interference of humans which makes it low cost.

Wearable sensors are smart devices that can be worn on body as a part of accessories [27]. They can be used in number of areas like security, entertainment, business, etc. It is helpful in providing security at home by providing reliable and accurate information. These devices are very tiny and lightweight so that it will be comfortable to wear 24/7. They are used for health monitoring such as blood pressure, temperature and heartbeat. BP can be easily measured by using these sensors rather than previous pressure cuff technique. For simple wearable wireless sensing device, representation is shown in Figure 1.29. Researchers have recently developed an artificial intelligence (AI) classifier that can detect a specific cardiovascular disease using a wearable wrist biosensor.

There are a number of sensors that can be used for health monitoring. Active sensor is the one that requires the external supply of power, e.g., photoconductive cell, while the passive sensor is that which does not require any external supply of power, e.g., electromechanical sensor.

Accelerometer: It is a device used to measure the acceleration along both axis as shown in Figure 1.30. Accelerometers are small in size and inexpensive devices that make it better to implement in real-life applications [28]. There exist a number of different types of accelerometers: a) piezo-resistive, b) piezoelectric and c) magneto-resistive, which measures the acceleration by using key concepts of technologies.

Fall detection, activity monitoring and immobility detection are the some examples of this sensor, which can be easily implemented in smart home

Figure 1.29 Wearable wrist biosensor to detect heart disease.

(a)　　　　　　(b)　　　　　　(c)

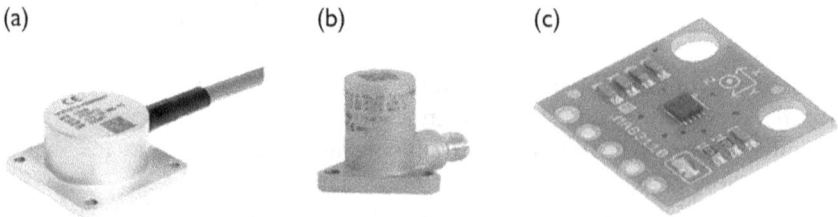

Figure 1.30 a) SE120 piezo-resistive accelerometer, b) CA134 piezoelectric accelerometer, c) MAG3110 three-axis magneto-resistive.

devices. Bio-sensors are those that are based on electrochemical technology. Image sensor is based on CMOS technology. This sensor is used in biometrics.

Motion detector: It is based on infrared, ultrasonic and microwave technology. When a person comes into the house and if he/she is not an authorized person, at that time this sensor detects the unusual activity. For smart home, the aim of this detector is to sense an unauthorized person and immediately send a text message to family members [29]. The aim of motion sensor is to detect the movement of a person and for this purpose, three techniques are used: a) passive infrared, b) ultrasonic and c) microwave. In this, passive infrared is used that detects body heat and used in smart homes for security purpose. Infrared radiation is generated if heat is generated by any type of object. This is the most common form used in homes, e.g., if a person enters into a room, this sensor automatically switch on the lights of the room. The other two technologies named as ultrasonic and microwave is a part of active motion sensors. They both emit sound signals and then reflection is measured for movement detection.

Telemedicine: It is a technology used for the transfer of important information, collected data for the treatment of medical situations or to provide health care facility at long sites by using audio and video techniques. Previously, the use of video channels was there for the proper interaction based on consultation services [30]. But now this technology has been expanded in a broader way. Now it can be used in number of areas such as artificial intelligence and patient education.

Radio frequency identification: It is a way of enhancing the process of handling the data as well as it is used for the storage and accessing the data by using radio frequency signals [31]. This technology can be used in various areas such as industries, security systems, manufacturing and retail.

There are three types of RFID tags: a) passive tags, b) semi-passive tags and c) active tags as shown in Figure 1.31. In this, tag means storage of that data that can be easily modified or accessed using radio frequency signals. The passive RFID tag does not require any power, and it uses external source for transmission of signals. Semi-passive tags also require an external source of power, but an active RFID tag requires an external battery.

1.7 INDIVIDUAL CARE IN PERVASIVE ENVIRONMENT

In a way to make home network more intelligent, smart devices are used with high processing and networking abilities. Networking ability means collection of sensors that can gather data and transmit that sensed data like temperature, humidity, light, etc. Here light means according to light in the home, curtains are moved automatically, temperature means to provide smart air conditioning system. An owner of the home wants to promote their homes to support these new technologies. It is a natural extension of

Figure 1.31 Types of RFID tags.

present human computing lifestyle. To make home smart is a way designed with new automatic technologies that provides facilities in a way to enhance safety, security, etc., of an individual in a way to maintain living independently by using smart kitchen, lighting system, etc. The basic and important function of developing a smart home is to know about the routine of persons and accordingly adapt with that routine. Number of sensors can be used for this technology like RFID, temperature sensor, pressure sensor, door sensor, etc. Technique that is used for wireless communication is infrared, radio frequency, ZigBee, Bluetooth, etc. Temperature sensor is used in kitchen to control the oven heat. In order to maintain activities of a person like standing, walking and unusual activities like fall detection, blood flow can be detected by number of sensors like impact sensor, accelerometer, etc. Light sensor allows light to be automatically on and off within a home. Photocell sensor allows the system to operate in dark environment. Emergency alarm can be used in smart home for emergency situation when a user wants immediate help. By using this device, message is sent automatically to caregivers. Individual care is an important factor to provide security and safety to an individual who is residing alone at home. So, to obtain this objective, first recognition of a person is done by using three modules: face recognition, voice recognition and similarity index. Initially, when a person comes and knock at the door, face of that person is to be recognized, if face is recognized properly, then go to next module, i.e., voice recognition. If voice of that person is also recognized, then access will be provided to that person who is at door by an individual who is alone at home. But, in case, if any one module is not recognized: in that case only third module is used, i.e., similarity index. It will check the percentage that how much percent voice of individual who is at door is matched with stored voices in the database. If similarity is greater than equal to 75%, then it is considered that user is recognized else declares unknown user. Last module is continuous monitoring the health parameters of that individual who is at home. Individual users can feel safer after knowing that they can contact with the people they know and can send message in case of emergency by using MATLAB® tool and smart phone. As in today's world, users have no time to visit doctor again and again, so patient requires on the spot monitoring and inform to their caregivers. Human health monitoring is emerging as a prominent application of sensor networks. Previously, the number of wireless sensors was placed on body of an individual and monitors the health parameters such as temperature, blood pressure and heart rate. These sensor devices are very tiny, light-weight and easy to carry.

The technology of smart home proves to be useful for elder persons or working women living alone in metros and for disabled persons living individually at home. In order to live individually at home in a healthier and safer way, elderly/disabled people or women can make use of latest features involved in smart home technology like monitoring system as shown in Figure 1.32, for fall detection or emergency system, etc.

A reliable emergency system as shown in Figure 1.33 must be there to monitor the condition of the elder and immediately call the doctor and the ambulance in case of emergency and notify the family members [32]. So, the objective is to provide a comfortable environment for all the persons living independently at home.

1.7.1 Neural networks

Artificial Neural Networks (ANNs) have been constructed with a structure similar to the human brain. The weight of an adult human brain weighs

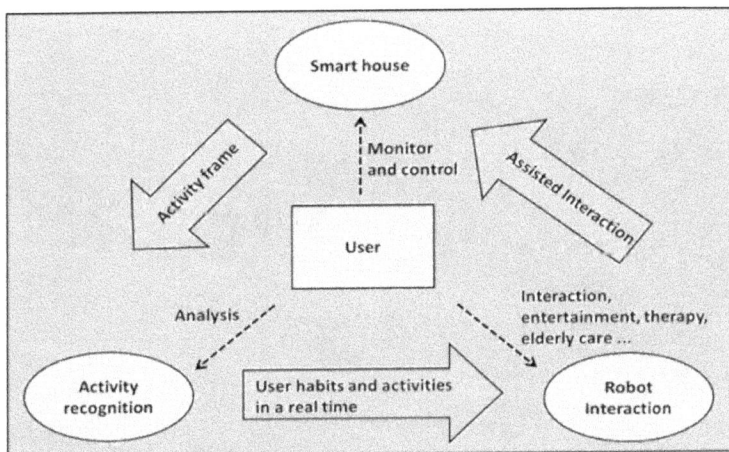

Figure 1.32 Activity monitoring system.

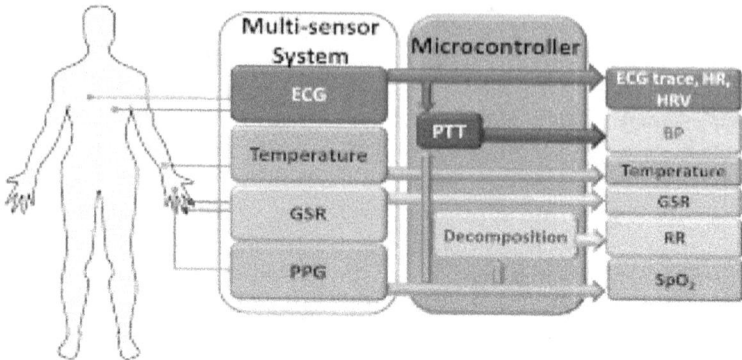

Figure 1.33 Reliable emergency system.

approximately 1.5 kg with a volume of 1260 cubic centimetres. The brain is composed of 200 billion neurons connected by 25 trillion synapses, glial cells and blood vessels. The neuron can be separated into three parts: cell body (soma), the dendrites and the axon as shown in Figure 1.34.

Every dendrite is contained with thin lines, which will receive signals from neurons surrounded by it. Each branch of the dendrite is connected to a single neuron with the help of a synapse [33]. NN is used to find patterns with their relationships in the data.

Figure 1.35 shows input neurons and output neurons with interconnected weights between them. The first step in this diagram is to take data or information with the help of input nodes in numeric expression form. ANN is a paradigm used to process the information by taking inspiration from a biological nervous system.

Flowchart of ANN in case of pattern recognition is shown in Figure 1.36.

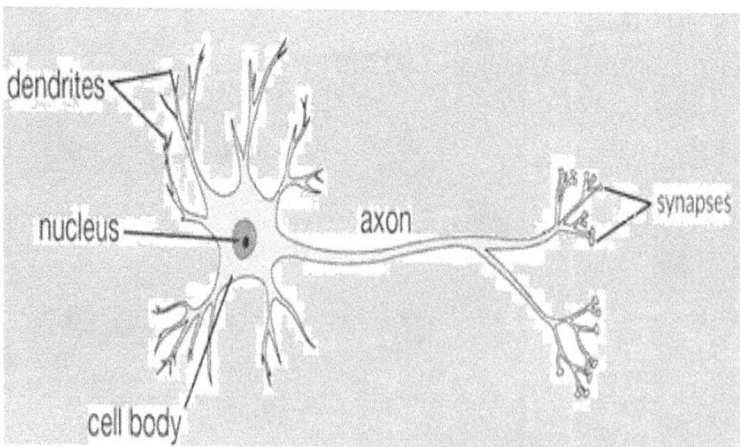

Figure 1.34 Structure of biological neurons.

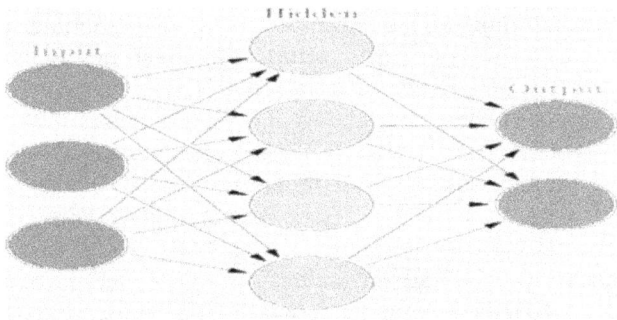

Figure 1.35 Structure of ANN.

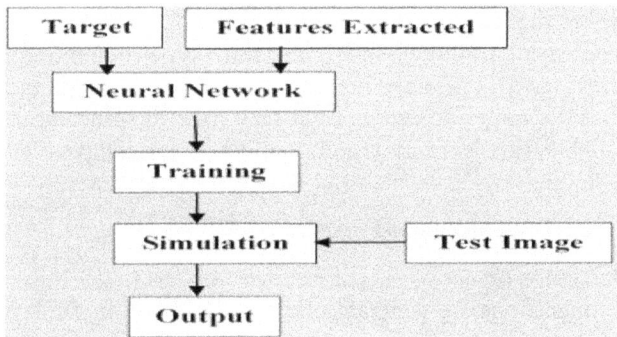

Figure 1.36 Flowchart of ANN.

Fundamentals of Neural Network

- **Processing units:** NN system is made up of number of processing units that represent neurons in a human brain. The units are divided into different groups like input units that receive the data to process, hidden units to manipulate and transform the inputted data and output unit to represent the final decision of NN [34].
- **Weights:** All the units of processing are connected in some way in NN. Every connection is having a value of weight on it. The input from a neuron is multiplied by the weight of connection which is sending the data, and output value is then send into the activation function of the network. If calculated value exceeds the threshold value of activation function, then the neuron will be activated else it will be inactive.
- **Computation method:** This method is used to determine a correct output by using the activation function. In this, neural network uses threshold logic unit to perform the computations. It takes input values from neurons then sum them and finally use activation function for the computation.

- **Supervised training:** This is the first method of training in which a network is provided with a series of number of inputs and then compares the targets with expected response or proper instruction is given out that will be desired output for a given input. The main issue in this is of error generation. It is desirable that generated output and desired output must be with minimum difference, e.g., Hebb net, Pattern Association memory, Back Propagation network and Counter propagation network.
- **Unsupervised or adaptive training:** The other method of training is called unsupervised training in which a network is provided with number of inputs but desired targets are not known. The grouping of input data items is maintained by the system. This is also called as self-organization or adaption method, e.g., self-organizing map, adaptive resonance theory, etc.
- **Reinforcement training:** This is the third type of training method in which a network is only presented with an alarm of indication of whether the target answer is right or wrong. If sufficient information is available, this learning readily handles a specific problem.

1.7.1.1 Network architectures

In the architecture of network, the neurons are arranged into number of layers and connections are generated between and within the layers. If two layers of interconnected weights are present, then it is found to have hidden layers, there are number of types of neural networks including vector quantization, hybrid networks, recurrent network, RBF networks, etc.

- Feedforward neural network

It is a network in which all the data will travel only in single direction. In this, there are two options: inputs are directly connected to outputs, it means there will be a single layer of weights or inputs are connected to the output layers using hidden layers. It consists of single-layer network as well as multilayer network as shown in Figure 1.37 [35].
Group behavior of all the neurons will describe the power of a network and on the basis of that it is assumed that this information will provide required output or not. There is no looping required for feedback. It is used for recognition of patterns.

- Single-layer network model

It means in single form where every input layer is directly joined to output layer but not in opposite form. It is also called as acyclic network. In this, input may be connected fully to the output units as shown in Figure 1.38.

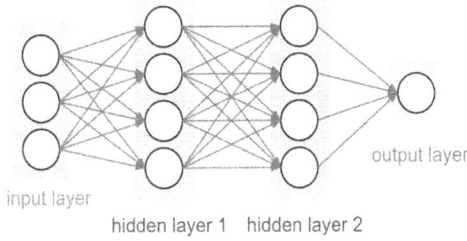

Figure 1.37 Feed forward neural networks.

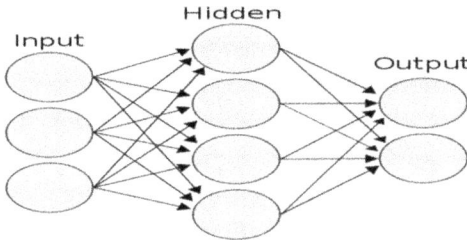

Figure 1.38 Single-layer feed forward NN.

- **Multilayer network model**

In this, the number of hidden units is more than one or it can be one where nodes are considered as hidden nodes. In this processing is done by first send the information from source node to first hidden node and second layer output will be used as a input to third layer and so on. The following picture 1.39 shows a multilayer network (Figure 1.39):

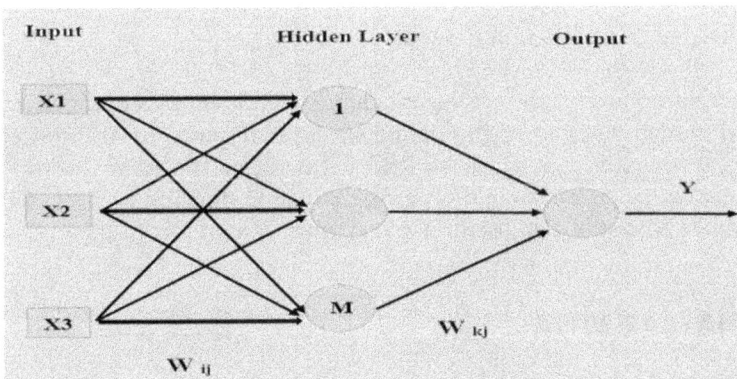

Figure 1.39 Multilayer feed forward NN.

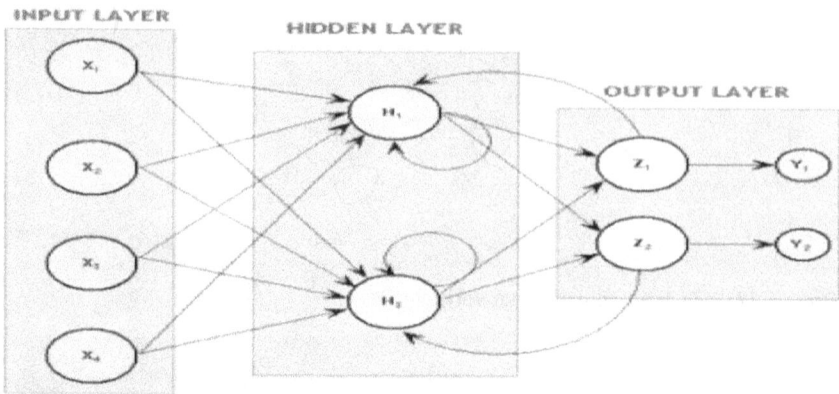

Figure 1.40 Recurrent neural network.

- **Recurrent network**

It is the simplest form of architecture in neural network. In this, all units of input layer are connected to all other units of output layer and every unit can act as both input as well as an output. It is that type of network in which number of hidden layers is greater than or equal to one with a feedback loop. In this, feedback means neurons output will be back to its input as shown in Figure 1.40.

- **Hopfield neural network**

This network is mainly used to recognize patterns and for that the network is trained with special algorithms. It is based on the concept of recurrent network. It consists of a set of neurons where each neuron represents a pixel of the difference image. It is also connected to all the neurons in the neighborhood as shown in Figure 1.41.

- **Radial basis function network**

As feedforward neural network is defined that it consists of input, output and hidden layer so in that if hidden layer depends on the concept of radial function it is called as RBF network, as shown in Figure 1.42. Gaussian point is used in this but its activation function is more complex than FFNN. It is also used for recognition of patterns [36].

1.8 MATLAB 2012A

MATLAB stands for Matrix Laboratory. There are two projects LINPACK and EISPACK used to provide simple process to matrix software.

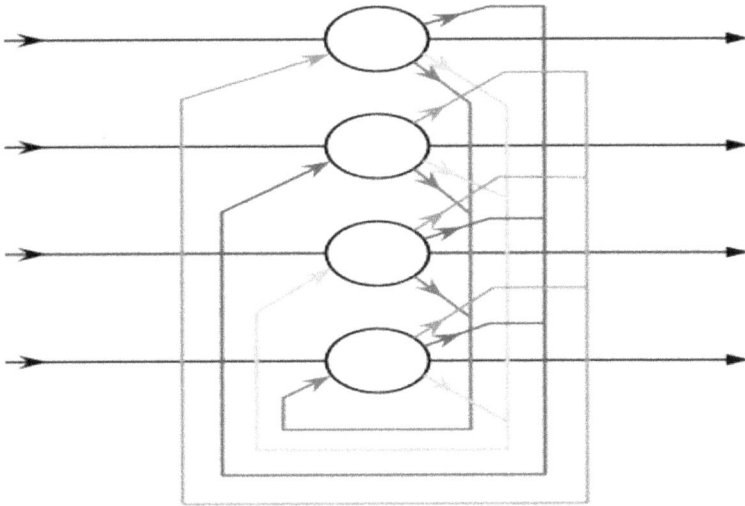

Figure 1.41 Hopfield neural network.

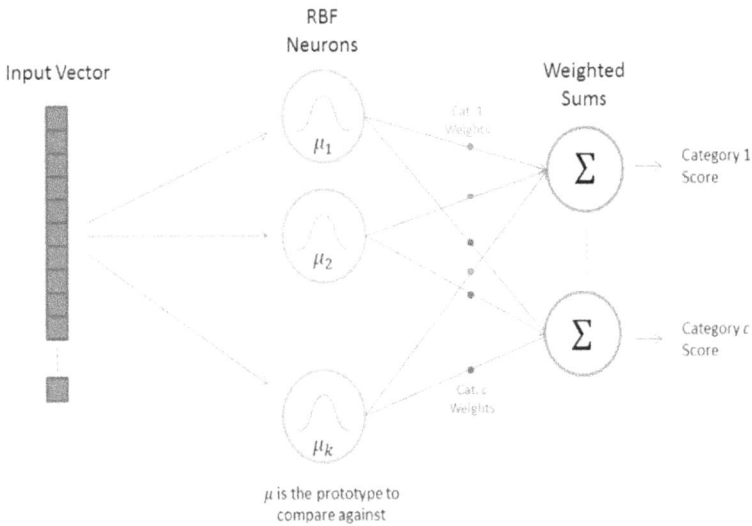

Figure 1.42 RBF neural network.

LINPACK means linear system package, and EISPACK means Eigen system package.

MATLAB has a number of built in routines that enables a wide variety of calculations and number of commands used for graphics that makes the availability of visualization results. A term is named as toolbox; it means number of applications collected in packages. There are different types of toolboxes used for simulation, signal processing and control theory with some fields engineering technology.

When a person logs in account, he/she enters in MATLAB tool by double clicking on shortcut icon of MATLAB on desktop. A window named as MATLAB desktop shows when MATLAB is started and this window includes other windows like command window, start button, workspace, command history, current directory and help button. When MATLAB started, the screen is like as shown in Figure 1.43. The tools and documents can be customized according to the needs of the users. Command Prompt (>>) is available in the command window [37].

Foe example, MATLAB is used as a calculator: A simple expression can be evaluated by using MATLAB, e.g., if numbers are calculated by this expression

$$2 + 2 * 3$$

It means after prompt (>>), like

$$> > 1 + 2 * 3$$

$$Ans = 7$$

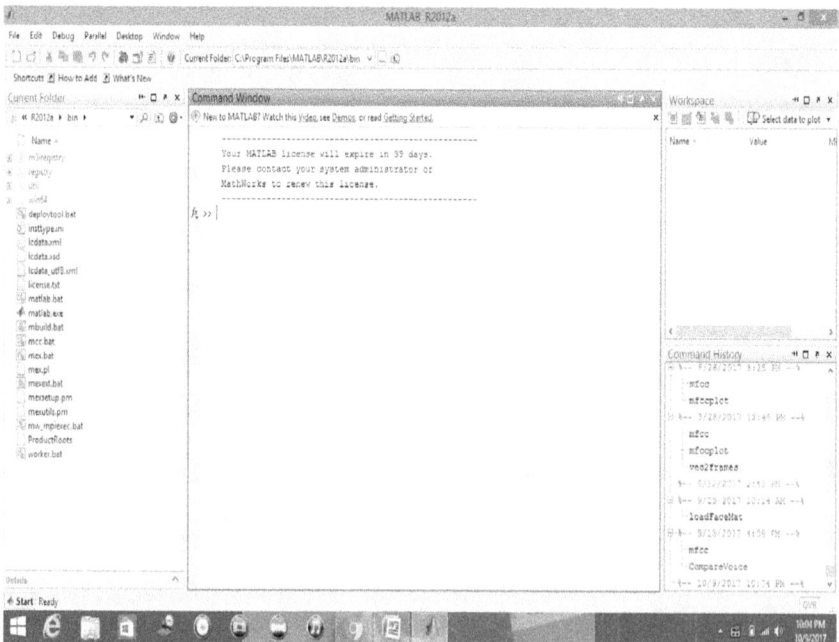

Figure 1.43 Initial screen of MATLAB®.

If output variable is not specified, in that case, default variable is the answer and its variable is specified like

> >X = 2 + 2 * 3

X = 8

Then the result will store in variable X. MATLAB session is ended by selecting File -> Exit MATLAB in main menu or type quit in the Command Window.

Getting help: MATLAB Help is used from Help menu in a way to view the documentation online. The best way of getting help is to use the Help browser by selecting this icon on desktop toolbar.

For example, if help is required for a particular function, i.e., sqrt, then type

Help sqrt

MATLAB plots: MATLAB tool is having a set of graphical tools. Few commands are used for results of computations or to plot a given data set. A simple plot is created by taking a vector of x-coordinates, x = {x1, x2, x3 ... xn} and a vector of y-coordinates, y = {y1, y2, y3 ... yn} locate the points (xi, yi) with i = 1, 2, 3 ... n and then join them by straight lines. To plot a graph, the command used in MATLAB is plot(x,y). For example, if vectors x = (2, 4, 6, 8, 10) and y = (3, −2, 3, 5, 2) produce the plot, as shown in Figure 1.44:

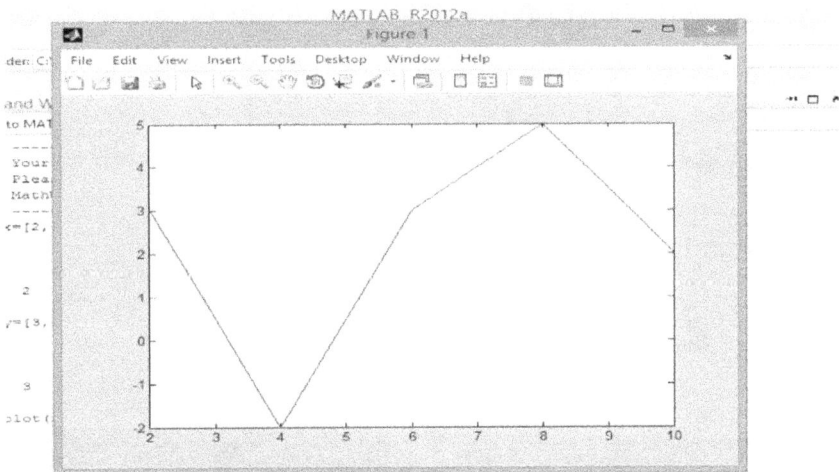

Figure 1.44 Plot for x and y.

>>x = [2, 4, 6, 8, 10]

>>y = [3, −2, 3, 5, 2]

>>plot(x, y)

MATLAB also helps in adding axis labels and titles. If x and y axes labels are added in above plot, it is by using these commands.

>>xlabel('x=0.2\pi')
>>ylabel('Sine of x')
>>title('Plot of Sine Function')

And graph is shown as in Figure 1.45. The graph of Sine Function with Titles is shown in Figure 1.46.

1.8.1 Matrix concept

The basic element of the MATLAB environment is matrices. A matrix is a two- dimensional array consisting of n rows and m columns. A number of operations are applied on matrices. First example is to enter a vector that is a basic case of a matrix. Its aim is to create a vector and matrices in MATLAB. Row vector is represented by 1*n dimension array, and column vector is represented by m*1 dimension array, as shown in Figure 1.47.

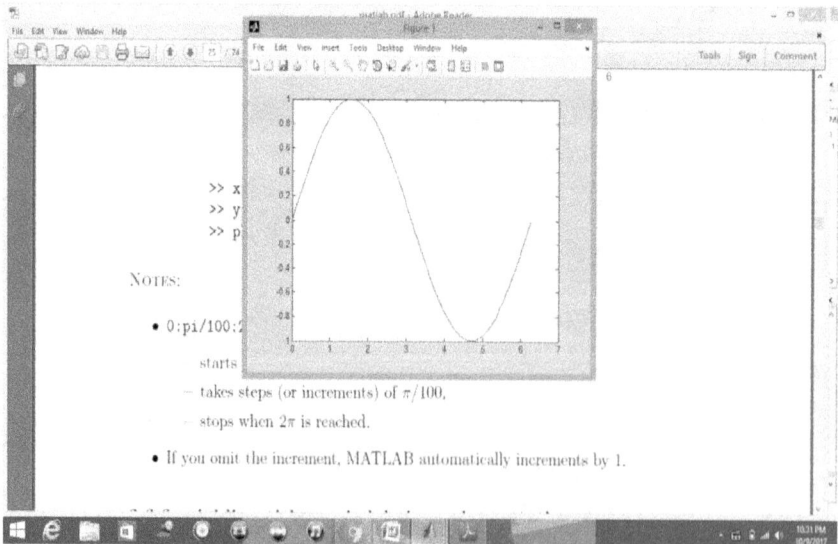

Figure 1.45 Plot for sine function.

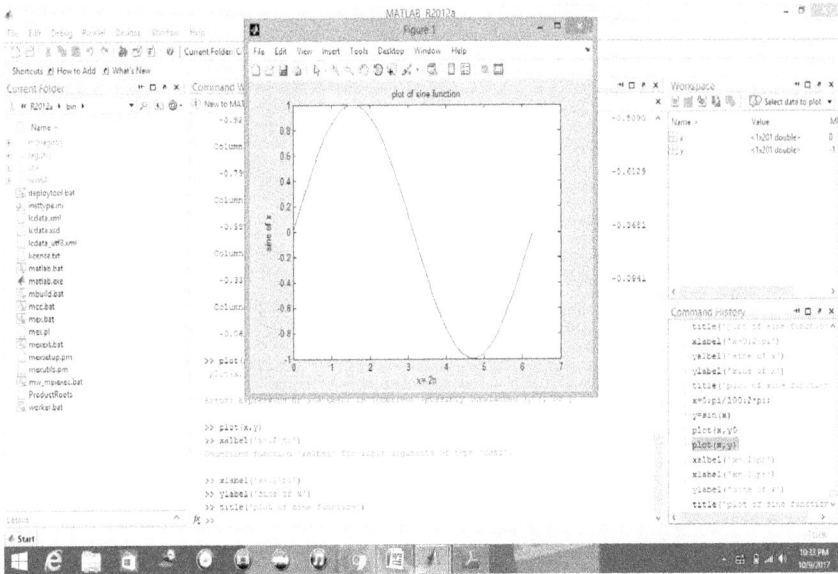

Figure 1.46 Plot for sine function with titles.

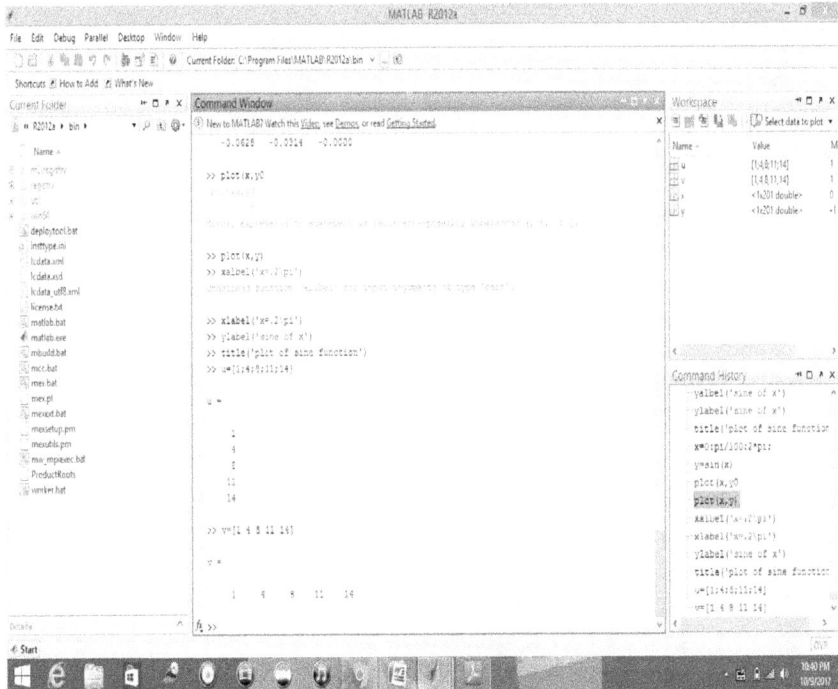

Figure 1.47 Representation of row vector.

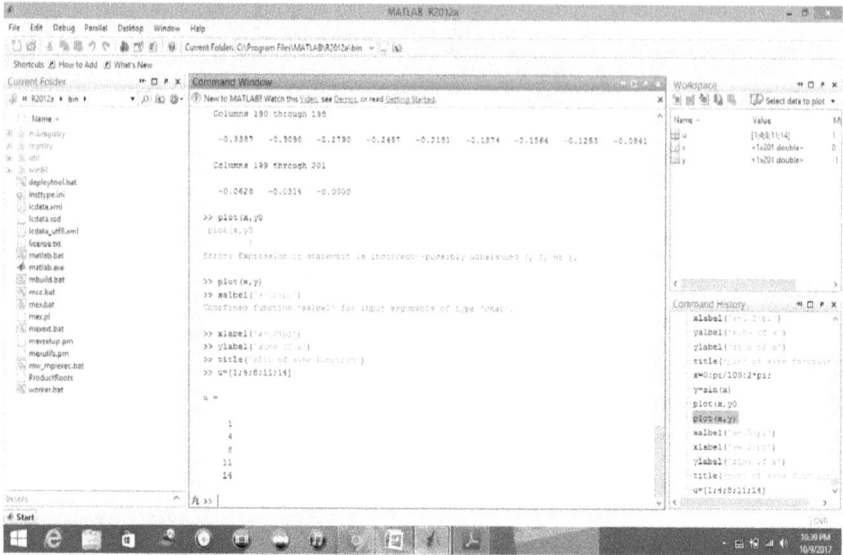

Figure 1.48 Representation of column vector.

For example, to enter a row vector v, type

>>v = [1 4 8 11 14]

Same as, column vectors are created in a way like, as shown in Figure 1.48.

>>u = [1; 4; 8; 11; 14]

A row vector can be easily converted into a column vector by using the transpose operator, which is denoted by an apostrophe or a single quote ('). For example, u=v' as depicts in Figure 1.49.

Matrix formation: A matrix is a collection of numbers. To form a matrix in MATLAB, start with-

Square bracket [
Elements are separated by space or comma (,)
Semicolon (;) is used to separate rows
Matrix ends with another bracket]
For example, to form a matrix as shown in Figure 1.50.
>> A = [2 4 6; 8 10 12; 14 16 18]

Result will be 3*3 matrix as shown in Figure 1.50.

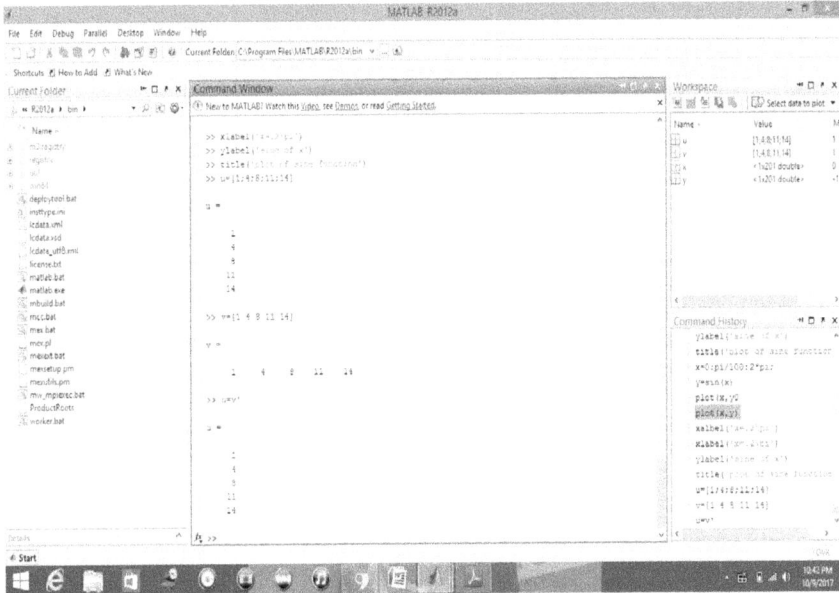

Figure 1.49 Representation of transpose operator.

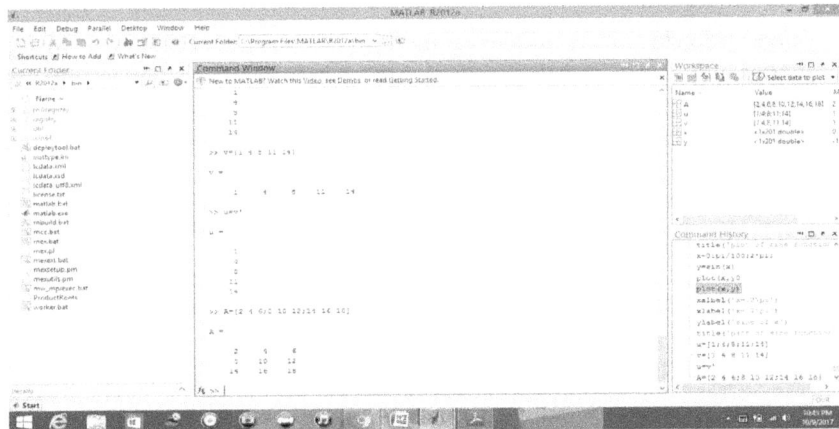

Figure 1.50 Representation of matrix.

Programming in MATLAB: As commands are executed in command window, the problem is these commands are not saved and executed again number of times. So, a different concept is used for executing repeated commands, it is-

- File creation with commands list
- Save the file
- Run the file

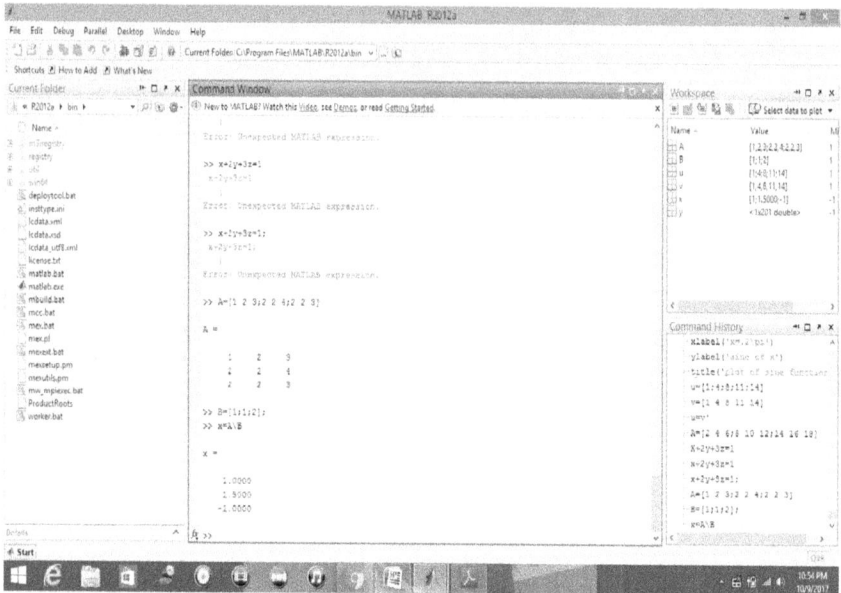

Figure 1.51 Representation of M-file.

There are two concepts defined in this part: M-file scripts and M-file functions.

M-file script: When MATLAB statements are contained in a sequence, that file is named as M-file scripts. It is having an extension.m, so it is called as M-files. User creates this file to store the code they write for a particular application. The representation of an M-file is shown in Figure 1.51. Consider three equations:

$$X + 2y + 3z = 1$$

$$2X + 2y + 4z = 1$$

$$2X + 2y = 3z = 2$$

Find x?

Use MATLAB editor for file creation.

File −>New −>M−File

Then enter the statements-

A = [1 2 3; 2 2 4; 2 2 3]
B = [1; 1; 2];
X = A\B

Then save this file, e.g., test.m by File -> save as -> location that suits. And run the file by select run or press run button or by F5.
Run by typing

>>test

M-file functions: Functions are the programs that accept the input arguments and return output arguments. Each M-file function is written as M-file or function in short.
E.g. function f = factorial (n)

% Factorial (N) returns the factorial of N
% Compute a factorial value
F = prod (1:n);

In this, first line starts with keyword 'function'. It gives function a name and number of arguments.

1.8.2 Image processing in MATLAB

MATLAB uses a toolbox for image processing. Images are composed of two- or three-dimensional matrix of pixels. Individual pixels contain a number or numbers that represent gray scale or color which is assigned to it. Color pictures are heavier in size as compared to gray scale pictures, so more power is required to process color pictures. In image processing, first step is to convert color pictures to gray scale pictures and then processing will be done on gray scale images. Before this, image is loaded to begin the processing. If image is colored, then first image is converted into gray scale for making processing simpler.

For conversion, a function is used, i.e., rgb2gray to change the color. There are two functions that are used in image processing: imread and imshow. Imread is used to read an image file with a particular format, and imshow function is used to show an image. Last is to write an image using imwrite function, it allows to save an image as any file supported by MATLAB, which are same as supported by imread.

1.8.3 Properties of image

- **Histogram:** It is a bar graph that shows data distribution. In image processing, it shows the number of pixel values present in an image.

Image can be manipulated to meet the specifications. For creating a histogram from an image, use 'imhist' function, 'histeq' function is used to for contrast enhancement 'graythresh' is performed for thresholding.

- **Negative:** It means the output image is the reversal of the input image. For 8-bit image, the pixels with a value of zero take on a new value of 255, while the pixels with a value of 255 take on a new value of 0. All pixels values in between take on similarly reversed new values. The new image appears as the output of the original image. This operation is performed by imadjust function.

- **Fourier transform:** The role of frequency is important to understand how image processing filters work. An image consists of collection of discrete signals so signals are associated with frequencies. There are two ways to know frequency content that it is low or high. If there is less change in gray scale values when image is scanned, it means there is low-frequency content contained within the image. If there is more change in the gray values when image is scanned, it means there is high frequency content contained within the image. Fourier transform is described as the transformation of a signal into its constituent sinusoids. For digital systems, if Fourier transform is designed in a way that discrete input values are entered by specifying sampling rate and computer generated discrete outputs, this is called as Discrete Fourier Transform. MATLAB uses a fast algorithm for DFT computations, which is called as FFT, whose command is fft.

- **Convolution:** In image processing, a linear filtering method is used named as convolution. It is the algebraic process of multiplying two polynomials. An image is an array of polynomials whose pixel values represent the coefficients of the polynomials. So, two images can be multiplied together to produce a new image with the help of convolution. In MATLAB, conv2 is used to perform a 2D convolution of two matrices.

- **Filters:** Filtering is used to modify an image, and it is used in image processing. There are mainly two filters used a) low-pass filters are those which blur high-frequency areas of images. This filter is used when unwanted noise is to be removed from an image. b) Median filters are those that can be helpful for noise removal from images. It is like an average filter, it examines the pixel and its neighbor's pixel values and return the mean of these pixel values. By this, noise can be removed but edges are not blurred as much.

1.8.4 Signal processing in MATLAB

The signal processing on MATLAB consists of number of tools for the computation. From generation of waveform to design of filters, all the operations are fully supported by signal processing toolboxes. The functions of signal processing are expressed in M-files to implement the tasks. Two built-in functions that are most commonly used in MATLAB are- filter and fft. Filter function is used on a sequence of data for filtering purpose, and fft is used to calculate the transform of sequence of data.

1.9 BOOK ORGANIZATION

The main objective of this research is to provide security to an individual who is living alone at home. The work has been organized in following chapters:

Chapter 1 introduces the basic concepts of pervasive computing with their challenges and issues. A number of biometric techniques are defined with face and voice recognition technology in detail, and different types of sensors are described in this chapter along with the objectives of this research work. A hybrid PICA algorithm is proposed to provide security to an individual at home is presented in Chapter 2. A hybrid MFRASTA algorithm is proposed for providing flexible trust-based security for the individual living independently is described in Chapter 3. In Chapter 4, a fingertip application has been proposed that works on camera lens of iPhone for health monitoring of users. Chapter 5 shows a GUI for complete implemented system and shows a comparison between proposed techniques with number of previously used techniques in terms of recognition rate. The conclusion of the proposed work has been presented in Chapter 6. Some points have also mentioned for future research to make this system more useful.

Chapter 2

Hybrid face recognition PICA algorithm for individual's security at home

2.1 INRODUCTION TO FACE RECOGNITION

The security framework assumes a crucial part in today's world. Thus, security occupations are made, and workers are recruited for this framework that runs effectively over some time. In any case, later on the world has been improved step by step investigating the new advances. Countless sorts of a robotized-based security framework start to be executed to diminish labor and cost [38]. The face is one of the most promising techniques of biometric technology, and it is the most common method of recognition. The human face conveys data about age, identity, gender, race and facial demeanors mirroring the feelings and mental states. Facial acknowledgment as opposed to a few other biometric qualities doesn't really require the collaboration of individual and can be acted in a subtle manner, making it especially reasonable for observation applications [39]. A problem with other biometric techniques based on fingerprint, iris, signature recognition, etc., are for collection of data, e.g., in case of fingerprint recognition [44], the user has to put finger in proper manner and direction. But in face recognition, collection of face images is not difficult and thus it can be used as a biometric technique. Each face has special features that define a particular user. Emotions can also be recognized by using pixel intensity, which corresponds features used for detection of emotions [45]. Face recognition is a biometric technique that is most frequently used in a number of applications such as banking, security information and virtual reality [45]. Image processing is a method used in face recognition to convert an image into digital form and perform some operations on that image so that we can extract useful information from it. It consists of three steps: take image using a scanner or camera, apply data compression or enhancements in order to analyze the data and take output in the form of altered image. The face recognition system involves three main steps: data acquisition, feature extraction and recognition [46] as shows in Figure 2.1.

Acquisition of data is the main step of the face recognition system, which includes the collection of facial images through a number of sources. The

DOI: 10.1201/9781003120933-2

```
──▶  Acquisition  ──▶   Feature      ──▶  Recognition  ──▶
                       Extraction
```

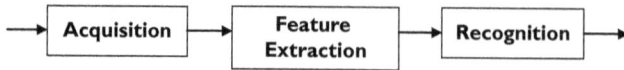

Figure 2.1 Steps of face recognition system.

number of collecting images may have different pose variations, illumination variations, etc. So, we can easily verify the performance of our recognition system with a number of conditions. Sometimes, the input image at the door is classified and different modification methods are applied to an input image. In this way, some undesirable parameters such as noise, lighting, blurred effects are removed and then it is used for next process [46]. Feature extraction is the next step of a face recognition system, which is used to take out the information from given images and after extraction, features are used in recognition step. In this step, dimensionality reduction is used with noise cleaning [47].

Recognition is the last step of face recognition systems used after feature selection and extraction, now the image is classified by using a number of classifiers. In this step, a number of images are taken and feature extraction is applied to them and stored in the database. When a new image comes for face recognition, it first performs face acquisition, feature extraction and then compares all the images in the database with this new image. Image is correlated with all the images stored in the database to identify an individual as shown in Figure 2.2 [47].

Every face recognition system consists mainly of two tasks:

- Verification
- Identification

In verification, one to one matching is applicable; as the system helps to match a particular individual with a specific biometric existing on file. As this system only compares the biometric files, that's why it is faster and efficient technique than identification, e.g., it seeks to provide the answers

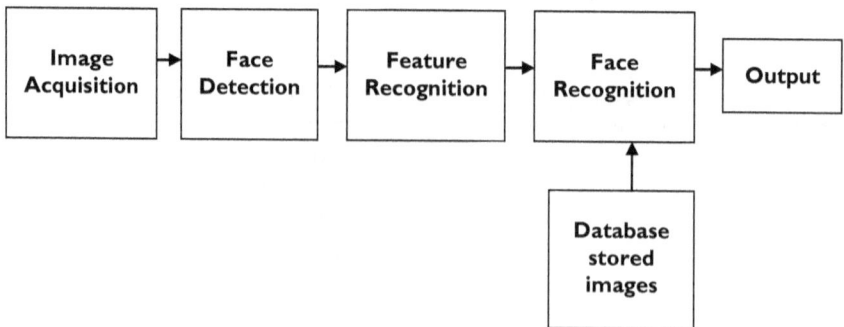

```
┌──────────┐  ┌──────────┐  ┌──────────┐  ┌──────────┐  ┌────────┐
│  Image   │─▶│  Face    │─▶│ Feature  │─▶│  Face    │─▶│ Output │
│Acquisition│  │Detection │  │Recognition│  │Recognition│  │        │
└──────────┘  └──────────┘  └──────────┘  └──────────┘  └────────┘
                                               ▲
                                          ┌──────────┐
                                          │ Database │
                                          │  stored  │
                                          │  images  │
                                          └──────────┘
```

Figure 2.2 Face recognition model.

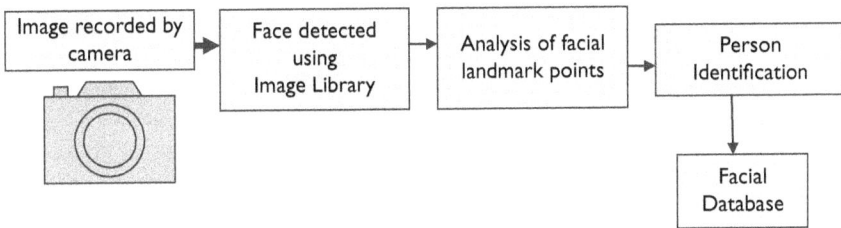

Figure 2.3 Basic steps of face recognition.

of specific questions like 'Is this person who is actually saying they are?' In identification, one too many matching is applicable, it means the image is identified in gallery database. It seeks to identify an unknown user, e.g., it tries to answer the question 'Who is this person?' The basic diagram of face recognition is shown in Figure 2.3.

There exist two types of face recognition methods: feature-based and holistic-based [48]. In feature-based method, features are extracted from the face such as eyes, lips and nose and geometric relationship between these features is calculated and finally reduced to a vector form. The method that is used to compute the distance between various facial features is a Euclidean distance method. Elastic bunch graph method is an example of this technique that is based on dynamic link structure. With this method, for every individual, a graph is generated which is used for recognition purpose by generating graphs for each new image in bunch graph method. After then, comparison will be performed by analyzing the new face image graph with all the graphs of face images stored in the database, and finally, similarity index is calculated.

Holistic-based method uses global representations for full image apart from local or some facial features. This method is again separated into two parts: statistical and artificial intelligence. In statistical approach, images are represented as two-dimensional arrays, and the main concept used is to calculate correlation parameter between the input image and the other stored database images. This method is very expensive as it does not work on variable conditions such as pose and lighting, and the second drawback is that it performs classification always in high-dimensional space. To overcome these problems, other methods can be used in order to gain more accurate and meaningful results, for example, PCA (Principle Component Analysis) algorithm which represents each face as a fusion of Eigen faces and any reconstruction is also done by using these Eigen faces.

In artificial intelligence approach [48], a number of tools are used to recognize faces like neural network and machine techniques. It means when principle components are extracted from PCA algorithm, then neural network is used as a classifier in order to reduce components in low dimensional space. SVM (Support Vector Machine) is the most commonly used algorithm in pattern classification. HMM (Hidden Markov Model) is also taken for face recognition task. The difficulty in using face recognition is it does not work under high variability in pose, illumination condition, etc. Problem with face recognition under varying pose and illumination condition was addressed that's why robustness is needed by using face geometry to cope with pose and illumination conditions. [48] So, a combination of PCA+ICA algorithm is defined in the proposed work for face recognition.

2.2 FACE RECOGNITION ALGORITHMS

Face is life's most important visual object that can be used for conveying identification. There are a number of algorithms used for face recognition.

2.2.1 Principle component analysis (PCA)

PCA is a measurable strategy that utilizes orthogonal change to change over a set of perceptions of possibly connected variables into a bunch of estimations of straightly uncorrelated factors are called principal components. The quantity of principal components is not exactly or equivalent to the quantity of original variables. This change is characterized so that the primary principal component has the biggest possible variance (i.e., represents, however, much of the changeability in the information as could be expected), and each succeeding part thusly has the most elevated difference possible under the limitation that it be orthogonal to (i.e., uncorrelated with) the former components [40]. Principal components are destined to be autonomous if the informational collection is jointly regularly distributed. PCA is touchy to the overall scaling of the original variables. Numerous papers of PCA are study here and find that, no papers are joined with the strategies of PCA and the significant model. That is the reason this sort of paper is composed where all these things are joined together which is straightforward and easy to understand for the beginner. PCA is a generally utilized mathematical tool for high-dimension data analysis. PCA has been utilized for face recognition, movement analysis and synthesis, bunching and for dimension reduction. It is a very popular algorithm developed in 1991 by Turk and Pentland [49], which is mainly used for reduction of dimensionality in compression-type problems. PCA originates from Eigenvector concept [49]. It is the most important tool used for analysis of data with pattern recognition that can be used in image processing. The linear transformation method is used by this algorithm in a way to map complete data from higher dimensional space into low-dimensional space. In this type of algorithm, real data set's shape and location will be changed if we transform it into different spaces, and it is used to calculate better discriminating components without the group's knowledge. It is a technique used statistically for analysis of data and for dimensionality reduction. The objective of this algorithm is to take data as a combination of a number of components [49]. A first principle component is a linear combination of original dimension along which the variance is maximized. A second principle component of original dimension along which variance is maximum and which is orthogonal to the first principle component. Nth principle component is a linear combination with highest variance subject to being orthogonal to n-1 principle components. To decrease the number of variables required for recognition of face, a statistical approach is used. Each and every image that is present in training set will be represented by a linear combination of weighted Eigen feature vectors, which are known as Eigen faces [50]. A term called as bias function is used that represents all the Eigen vectors, which are generated by a covariance matrix of the given training set. In the recognition phase, a new image is projected onto subspace and Euclidean method is used to calculate distances amongst various facial features for classification purpose. PCA is also called as Karhunen Loeve transformation [50]. It is used to remove dimension and correlation problem and analyzes the data patterns by expressing the similarities and differences [50].

The steps taken to implement PCA algorithm are as follows:

All the facial images that have been taken are in two dimension says N*N. The objective is to extract the principal components (Eigen faces) that can represent the whole faces stored in a training set [51]. Assume the number of images in training set represented by M and linear form of every image in training sets are represented by I_n.

- Convert images of the training set to image vectors so we map N*N into $1*N^2$ vector.
- Normalize face vectors, it means to remove all common features that share together from each face so that each face can be left behind with only its unique features.
- Calculate mean of all face vectors

$$Mean = 1/(M) \sum_{i=1}^{M} I_i.$$

- Subtract mean from each face vector I_i so that normalized face vector is left.

$$K_i = I_i - Mean$$

- Calculate Eigenvectors from the covariance matrix.

$$C = 1/(M) \sum_{i=1}^{M} Ki * Ki^T = B * B^T (N^2 * N^2 matrix)$$

where $B = [K_1 \ K_2 ... K_M]^T$ (N^2*M matrix)

- Calculate Eigenvectors from reduced covariance matrix C or $B*B^T$, let it be u_i but $B*B^T$ has a very large size and computation of Eigenvectors for it is not possible. So, instead find out the Eigenvectors for matrix B^T*B. Let v_i be the Eigen vectors.

$$B^T * Bv_i = I_i * v_i$$

Relationship between v_i and u_i

$$B^T * Bv_i = I_i * v_i$$
$$B^T * B * Bv_i = I_i * Av_i$$
$$C * Bv_i = I_i * Bv_i$$
$$Cu_i = I_i u_i$$

where $u_i = Bv_i$
- Select best Eigenfaces that can represent the whole training set.
- Convert low-dimensional Eigenvectors to original dimensionality.
- Represent each face image a linear combination of all Eigenvectors. It means each face in training set can be represented as a weighted sum of Eigenfaces and mean face.
- Recognizing an unknown face. To recognize a face image, steps which are followed:
  a. First convert face image into vector form I.
  b. Normalize it by subtracting the mean from it.
     K = I-mean
  c. Project normalized face vector (K) onto Eigen space in order to obtain a vector W.

$$W = [w_1 w_2 \ldots .. w_N]^T$$

where $w_i = u_i^T * K$
  d. Find e_r, it gives minimum distance the given face has from another face belongs to training set and the given face belongs to that person to whom the face in the training set belongs.

$$E_r = \min \|w - w^T\|$$

  e. If $E_r >$ threshold, the user is unknown else user is known.

Figure 2.4 describes the steps taken in PCA algorithm to take out the principal components of the face image [51]. In the PCA algorithm, all the images must be of equal size. The normalization process is applied to normalize the data for all images. Every image is considered as a single

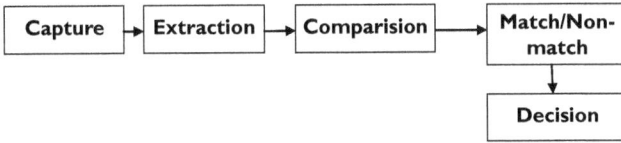

Figure 2.4 Schematic diagram of a face recognizer by PCA.

vector, it means all the face images are kept in a single matrix and every row of the matrix shows an image. Next step is to calculate Eigenvectors for these images, these Eigenvectors are also called as Eigenfaces. Every image is a linear combination of weighted sum of these Eigenfaces. Whenever a new image is registered for recognition, Eigenvectors for that image is first created and then the distance between new face image which is at the door and all the stored images in the database is computed. The main goal of using PCA is to get important features of data by using a method called matrix method. PCA Eigenfaces method does not work on images directly, so the matrix method is used which converts all the images into vector form and stored them in a single matrix. Figure 2.5 shows a schematic diagram of a PCA and neural network-based face recognizer.

2.2.2 Independent component analysis (ICA)

It is a statically and computational strategy for finding basic factors or parts from multidimensional statistical information. ICA is different from other methods in a way that it searches for components, which are having both characteristics: statistically independent and non-Gaussian. ICA characterizes a generative model for the noticed multivariate information, which is ordinarily given as an enormous collection of samples. In the model, the data variables are assumed to be linear mixtures of some unknown latent variables, and the mixing system is also unknown. The latent variables are assumed non-Gaussian and mutually independent, and they are called the independent components of the observed data. These independent components, also called sources or factors, can be found by ICA. The data analyzed by ICA could originate from many different kinds of application fields, including digital images, document databases, economic indicators and psychometric measurements [41]. ICA is used to analyze larger order data or we can say multivariable data [52]. The main technology used in this algorithm is conversion of higher dimensional space into low-dimensional space in such a way that new space which is contained converted variables will define the essential features of the data. Cluster analysis is a technique that is used in this algorithm [53]. A cluster is represented as a group of a number of data elements and cluster analysis is defined as a technique used for allocating space in a region to data where large concentration is present and that region is called as cluster [53]. This

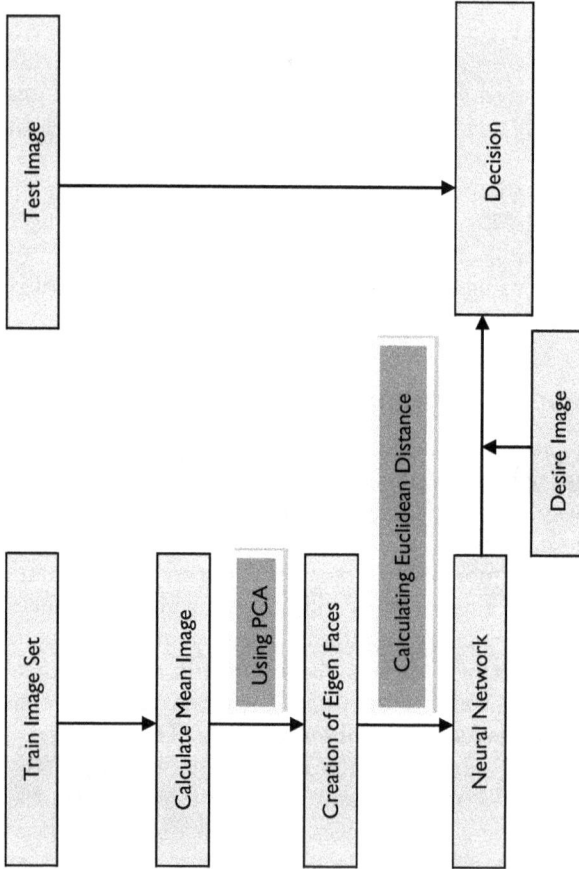

Figure 2.5 Schematic diagram of a PCA and neural network-based face recognizer.

technique is basically used for blind source separation problem that is also called cocktail party problem. This problem is defined as if there are three people speaking simultaneously in a room by using three microphones that are placed at different locations in order to pick sound produced by the microphone and each microphone records a mixture of all the voices in the room. The objective is to separate the voices of the individual users using only the recorded mixtures of their voices. Independent Component Analysis (ICA) is a method of addressing this problem of blind source separation [54]. It is a generalization of the PCA algorithm which distinguishes between higher order and second order moments of input. Its main work is to divide a signal in a number of combinations of independent signals.

In this, an algorithm that is used is info max [55] proposed by Bell and Sejmouski for separating the analytical independent components of a data set.

Steps of ICA Algorithm are:

Every face image is a unique linear combination of these independent components.

$$R = A * U$$

where R represents images of faces, A defines the unknown mixing matrix and U represents statistically independent components.

- Perform sphering on the training set. It includes centering and whitening process applied on the data set.
  a. Centering: In this, row means are subtracted from the data set and then data set is passed through a whitening filter.
  b. Whitening: It is calculated by a formula:

  $$W_z = U * E^{-1/2}$$

  where U and E are Eigen vectors and Eigen values, respectively. This step removes the stats of data, both covariance and mean are set to zero and variances are equalized.

- Calculate $W - W * W_z$,
- Calculate basis image.

$$B = R * W_i^{-1}$$

- Project test image into Eigenfaces.
- Compute coefficient

$$B_{test} = R_{test} * W_i^{-1}$$

Both PCA and ICA algorithms are used to reduce measured components into a smaller set of components. PCA is adequate if the data are Gaussian, linear and stationary. If not, then ICA comes into the role. In simple words, PCA helps to compress the data, and ICA helps to separate the data [42]. Figure 2.6 shows a system that needs an input image. By inputting the image, it will find out the Eigen values for all training images. After then, a test image is selected and apply ICA algorithm onto the considered test image. Finally, from the considered test image, mean image is calculated. If it is matching with the training image, in that case, person is known else no result found. Figure 2.6 shows a block diagram of ICA based face recognizer.

2.2.3 Linear discriminant analysis (LDA)

LDA is a dimensionality reduction technique that is used like a pre-processing step for face recognition. Its aim is to project the features from high-dimensional to low-dimensional space for avoiding the dimensionality curse and most importantly reduce cost as well as resources. LDA technique was introduced by Ronald A. Fisher with name Linear Discriminant or

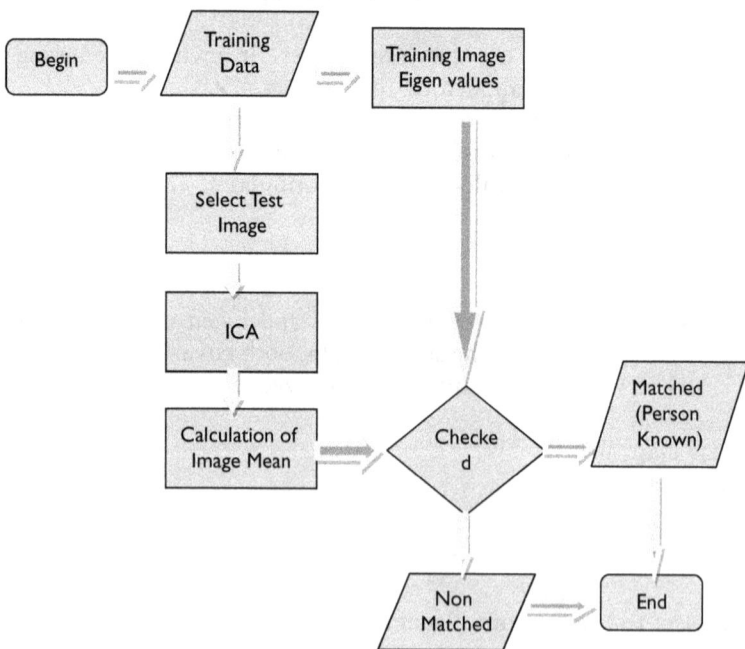

Figure 2.6 Block diagram of ICA-based face recognizer.

FDA (Fisher Discriminant Analysis) in 1936. This is one of the classical techniques for recognition of patterns, where this technique is used for finding the features of linear combination which separates object classes. The combination of results can be used like a linear classifier. Fisher Discriminant analysis is also a part of the linear discriminant analysis technique. In this analysis, a variety of pixel values can be used to represent each face. LDA is used to create a more manageable form by reducing number of features before classification happens. It is used to search for linear type of transformation in such a way that clusters of features will be very separable, which can be achieved by analysis of scatter matrix. This method is basically used to handle that case in which intra class frequencies will not be equal and we can calculate their performance by using any random data set. It maximizes the ratio of inter class variance with intra class variance for any data set by giving guaranteed separability [55]. This algorithm maximizes the between class scatter matrix, while minimizing the within-class scatter matrix by using a transformation as shown in Figure 2.7 [43]

The aim of using LDA is an improvement over PCA using the class membership function of training images taken into account when finding for a subspace. It is based on the concept that single face image is taken as input and uses score function with the Mahalanobis distance method concept [55]. Its advantage is that it maximizes the distance between clusters but minimize the within cluster distance, but it does not consider multiple images as input.

Algorithm:

It is very similar with the algorithm of Eigenface, in this first; training images are projected onto a subspace. After then, a similarity measure is

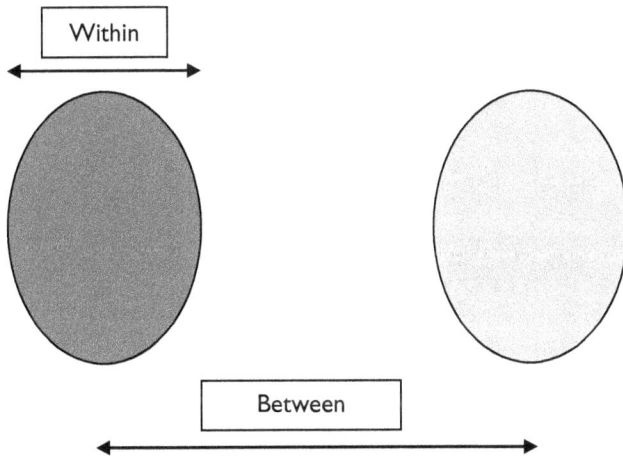

Figure 2.7 Class separations in LDA.

calculated by projecting a test image onto the same subspace. The measurement of subspace is measured with steps defined as:

1. Calculate variation within class by a measurement of the amount of variation between items belongs to same class.
2. Calculate variations between classes.
3. To compute Eigen vectors and Eigenvalues for both the categories: variations within class and between classes.
4. Then, sort the Eigenvectors by their attached Eigenvalues from high to low and keep the highest value Eigenvectors. These Eigenvectors are the Fisher basis vector.
5. Calculate the dot product of images with the fisher basis vector. This calculation projects images onto a subspace. Once trained images are projected onto a subspace, they must be collected and categorized as how close they are to one another. Computing the distance between images can establish their similarity. For example, images with the smallest distance between them can be considered close and similar, whereas images whose distances are farther apart would be considered dissimilar.

LDA technique tries for the discrimination of data set by using dimensionality reduction method, and the block diagram of testing phase of LDA is shown in Figure 2.8 [43]

2.2.4 Discrete cosine transform (DCT)

This is one of the exact and robust recognition systems and by using different normalization methods; its robustness to varieties in geometry of face

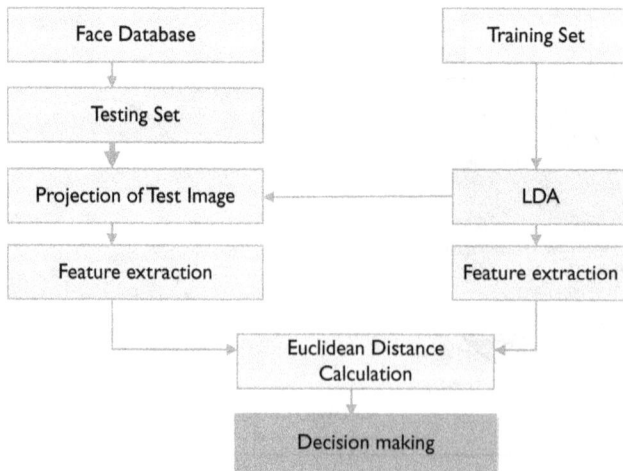

Figure 2.8 Block diagram of testing phase of LDA.

and illumination can be increased. Property of energy compaction enables DCT to be used in image processing and analysis of the signal. DCT easily compresses most of the signal information in number of coefficients and this is the reason that it is used for feature extraction of facial features. When Discrete Cosine Transform is applied on face image, it will provide a matrix containing low- and high-frequency coefficients [56]. Performance of DCT is largely affected by altering the coefficient magnitude at the top left corner of the matrix. In this, before applying feature extraction, illumination normalization is applied in order to compensate these coefficient variances and set the blocks to more equalized intensity. DCT is a very efficient technique used in image coding. In order to compensate with image quality, some coefficients that are having high frequency can be ignored. In case of variation in facial expression, illumination conditions, etc., two-dimensional DCT methods are used for feature extraction in order to increase the performance of the recognition system. The challenge of DCT technique is uncontrolled illumination conditions for extraction of features in real world applications. Illumination normalization (IN) is a preprocessing technique that compensates for different lighting conditions. It consists of a number of data points waving at different frequencies are represented in terms of addition of cosine functions [56]. A transform is defined as an operation that is used in signal conversion, i.e., to convert one domain into another domain by use of inverse transform. This transform is used in compression standard in which system receives face as an input image, then normalizes or crop that face, calculates distance between faces and finally comparison is done. DCT also attempts to decorrelate the image data as other transforms. After decorrelation, each transform coefficient can be encoded independently without losing compression efficiency. It works well when there is less variation. It works well under same orientation, pose, illumination, etc., but its success rate is not high.

2.2.5 Gabor wavelet

It characterizes the image as localized orientation selective and frequency selective features. A filter is used in this technique to extract the local features, which are actually applied on number of images for feature extraction aligned at a particular angle or orientation. [44] This technique is named after Dennis Gabor [57]. It is a band pass linear filter that works well in large variation in pose and illumination conditions. It achieves an optimal resolution in both spatial and frequency domains. Wavelets are functions that satisfy certain mathematical requirements and are used in representing data or other functions. Wavelet technique is used to process data at different scales or resolutions. If a signal is having a large window, then gross features are calculated and if a signal is having a small window, in that case small features are calculated. Wavelets are well suited for approximating data with sharp discontinuities. This technique can be used in

```
                                    ┌──────────────────────┐
                                    │     Face Dataset      │
                                    └──────────────────────┘
                                              ↓
                                    ┌──────────────────────┐
                                    │  Gabor Filter Creation │
                                    └──────────────────────┘
                                              ↓
                                    ┌──────────────────────┐
                                    │   Feature Extraction   │
                                    └──────────────────────┘
                                              ↓
                                    ┌──────────────────────┐
                                    │ Fusion applied on features │
                                    └──────────────────────┘
                                              ↓
        ┌──────────────────┐        ┌──────────────────────┐
        │  Testing Dataset  │        │ To apply Neural Network │
        └──────────────────┘        └──────────────────────┘
                 ↓                            ↓
        ┌──────────────────┐        ┌──────────────────────┐
        │ Feature Extraction│ ·····> │    Classification      │
        └──────────────────┘        └──────────────────────┘
                                              ↓
                                    ┌──────────────────────┐
                                    │   Result Evaluation    │
                                    └──────────────────────┘
                                              ↓
                                    ┌──────────────────────┐
                                    │ Comparison and Decision │
                                    └──────────────────────┘
```

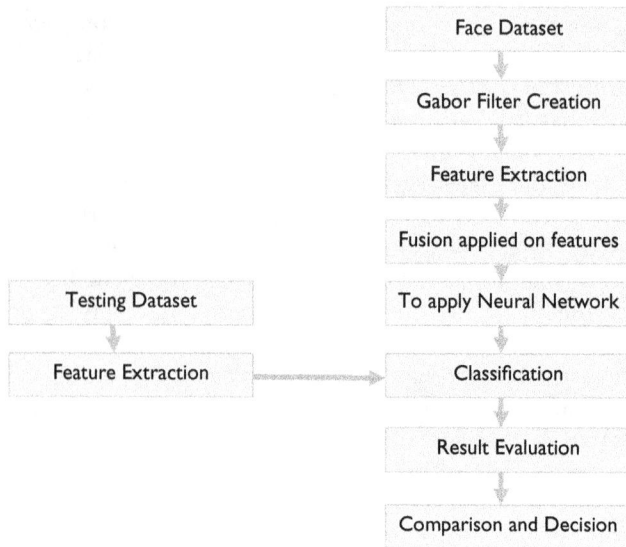

Figure 2.9 Block diagram of gabor wavelet techniques.

a number of areas: astronomy, acoustics, nuclear engineering, signal and image processing, music, magnetic resonance imaging, speech discrimination, optics, fractals, turbulence, earthquake prediction, radar and human vision. Figure 2.9 shows a block diagram of Gabor Wavelet technique.

There are number of features which make this technique attractive for use in number of applications:

- Illumination variation insensitivity
- Change in facial expression insensitivity
- Shift and rotation invariance

With these benefits, there are also some drawbacks in this technique:

- High computation complexity
- Requirement of memory capacity
- Dimensions of feature vectors are very high

PCA with ICA algorithms work well in less or more pose, illumination variation, etc., and for calculating the distance between facial features, they use Euclidean distance concept, dimensionality reduction method is used by PCA algorithm which is necessary for compressing the data while retaining useful information. Extraction of independent features is done by ICA algorithm, but other algorithms do not work well in less or more pose variation, illumination variation, etc. LDA algorithm does not consider multiple images at a time and based on Mahalanobis method for the distance calculation [57]. DCT works well only in when there is less variation and Wavelet works well only in the large variation of pose and illumination factors.

2.3 PROPOSED HYBRID PICA ALGORITHM

Automatic face recognition has been taken a highly active area of research and health monitoring of a person living independently at the home in an automatic way is a method which is used to give health-related information of individual persons to their family members. Figure 2.10 provides an integrated model for face recognition.

A combined approach of PCA + ICA is used for feature extraction in our work as this approach helps in obtaining our objective with high recognition rate as compared to individual feature extraction techniques and with this combination; a neural network named as back propagation is used as a classifier. The evolution of PCA [58] is done by using Eigenvectors and with the help of these Eigenvectors, features are considered for a given input, which is called as principle components. Its advantage is a dimensionality reduction process that includes the data compression, but there is no loss of information.

ICA is a technique that is very useful [58]. A feedforward neural network is a simple network consists of three layers: a) the input layer, b) hidden layer and c) output layer.

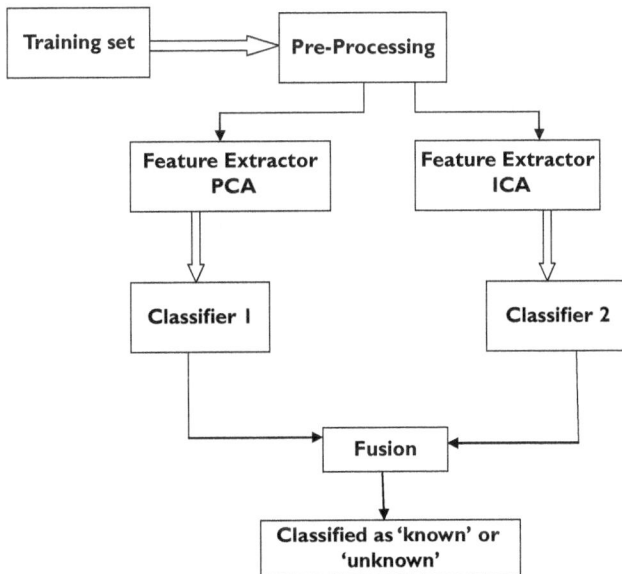

Figure 2.10 Integrated PICA model for face recognition.

Most of the work done in the face recognition system previously consists of single feature extraction technique with classifier, but in the proposed work four step procedure has been used.

    a. A training set is taken and its preprocessing is done in which images are normalized using different methods such as PCA and DCT.

    b. Independent features are extracted by using two face recognition algorithms PCA + ICA in combination.

    c. Feedforward neural network is chosen for classification that classifies a given input image.

    d. The output of each neural network is combined to classify input image that defines the category of that input image.

Proposed Hybrid PICA Algorithm:

1. First a training set is established.
2. Then, mark features function on training set is applied to mark eye and mouth position of the input image.
3. Apply align faces function to convert the face images in pgm format from JPG format as pgm format is supported by MATLAB® tool.
4. A matrix named as C is used to store all the images.
5. To apply the PCA function:
   - To center the data: Every trained image is centered by taking each of the training images subtracted from the mean image and stored in a matrix named as A.
   - To calculate covariance matrix: Centered images in the matrix 'A' are multiplied by its transpose to form a Covariance matrix L. The size of this matrix is $N^2 * N^2$, if each image is of size N*N.

$$L = A * A'$$

   - To calculate Eigenvectors: Finding Eigenvectors of $N^2 * N^2$ matrix are intractable. So, use of matrix A'*A is easy to calculate Eigenvectors.

$$C = 1/(M) \sum_{i=1}^{M} Ki * Ki^T = B * B^T (N^2 * N^2 matrix)$$

6. To apply ICA algorithm: From previously calculated Eigenvectors, ICA is analyzed with the help of centering and learning.
   - Centering: It means sphering of x or description as to remove first- and second- order stats from x.
   - Learning: To calculate Xx that holds original data.

$$Xx = inv(wz) * x$$

- To calculate output covariance: It is calculated by using the mean extracted and original data.

$Cov\,(xx')$

where $uu = w * wz * xx$
- The threshold value is set and results are shown.

7. Send text message to category members using MATLAB tools and a 3G dongle.

2.4 IMPLEMENTATION

To measure the performance of PCA and ICA, a code for the hybrid PICA algorithm is generated using a MATLAB tool. This tool is preferred in order to increase the recognition rate as compared to implementation in C or Java language. A combination of principal component analysis (PCA) with independent component analysis face recognition algorithms [58] is used for facial feature extraction and as a classifier or to recognize a face, feedforward neural network [57], [58] is used.

PCA implementation: PCA algorithm consists of two functions: PcaFn.m and Test.m. PcaFn.m is responsible for initialization of algorithm, training of data and projection of images on low-dimensional subspace.

In this, input contains the path of data set of all preprocessed images and path of images in training list. The output contains data in a matrix form, mean of images, Eigenvalues and low-dimensional subspace.

Steps of PCA implementation:

- Read face images from a path, reshape them into vector form and creates a matrix C.
- Create a training data list, which contains images from the data set.
- Calculate mean of images.
- Subtract mean from all the images and save the results.
- Calculate Eigenvectors and Eigenvalues.
- Calculate Eigenfaces.
- Projection of face images onto low-dimensional subspace.

Test.m is responsible for image recognition. It reads the image and calculates the difference between image to be tested and all the other images stored in the database. Image with shortest distance is considered as a recognized image.

In this function, input contains the path of the image, mean of image, low-dimensional subspace and output provides a matching score and associated image from database.

ICA implementation–ICA algorithm consists of five functions: PcabigFn.m, spherex, zeromn, runica and sep96. PcabigFn.m is responsible to compute PCA by calculating covariance matrix and reconstructing Eigen vectors of the smaller covariance matrix.

In this algorithm, input contains a matrix C with faces images presented in columns of the matrix and output contains V, R, E. V means a matrix containing Eigenvectors in the columns, R means PCA coefficients in its row and E means a matrix that contains Eigenvalues associated with Eigenvectors.

Runica is used to run an algorithm named as info max in order to find n independent components, where n represents the number of input face images provided to the algorithm. In this, input contains a class having a matrix x containing gray values of images in rows and output contains w & uu, where w means a matrix containing independent components and uu means separated output signals.

Spherex is used by runica function. It spheres the training vector x by passing it through a whitening matrix w_z. In this, input contains a function required x to be predefined and output contains w_z & x, where w_z is a whitening matrix of.

Sep96 is run by runica function. This implements the learning rule of info max algorithm. It spheres the training vector x by passing it through whitening matrix w_z. In this function, input contains x, w, count, sweep, where x defines the gray values of the images in its rows, w defines a matrix that hold the weights from learning rule initialized to identity matrix, count explains a counter increase during iterations and sweep holds a counter for number of sweeps and output holds the value of w which contains a matrix containing training weights calculated by an algorithm.

Complete steps that how a face can be recognized and when the face is recognized as known or unknown, a procedure of sending messages on the cell phone of family members using MATLAB [59] tool and the dongle are

Figure 2.11 shows the basic GUI of the face recognition system. It includes all the steps of recognizing a face like mark features, align features, load faces, testing and training.

Mark facial features: This is the first step in recognizing a face. It is included in preprocessing function which after applying takes images of size 512*512 pixels in JPG format. In this, double features of right eye, left eye and mouth are marked manually as shown in Figures 2.12 and 2.13 and get a matrix Labels.mat of size 1*6 which is showing value of six features for single face in Figure 2.14. This figure shows a matrix form having 6 columns used for extraction of features by using ICA algorithm and the features that are left are then extracted using the PCA algorithm. In the proposed work, total 30 features are extracted for face recognition.

Align Faces: This is the second step after mark features. In this module, mouth and eye positions are aligned into a pgm format from JPG format as shown in Figures 2.15 and 2.16. When JPG format is used in mark features

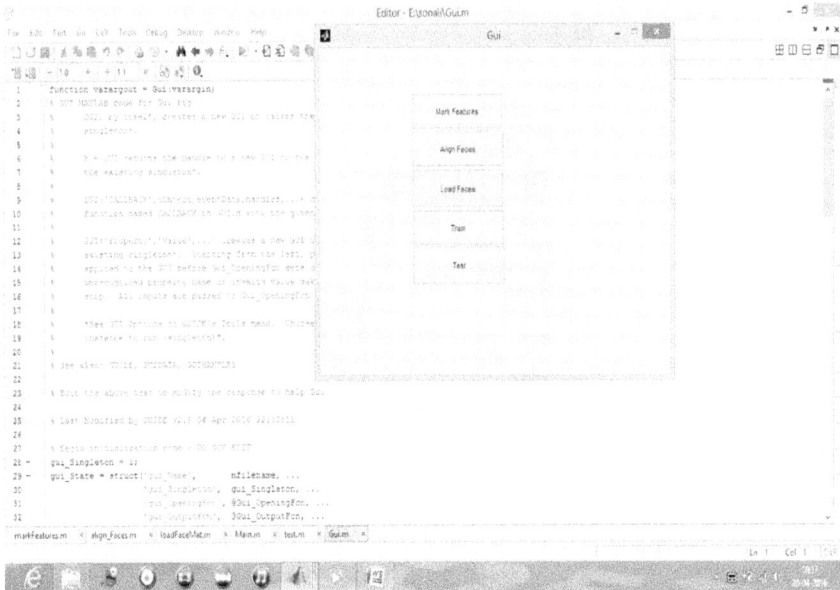

Figure 2.11 Interface of face recognition process.

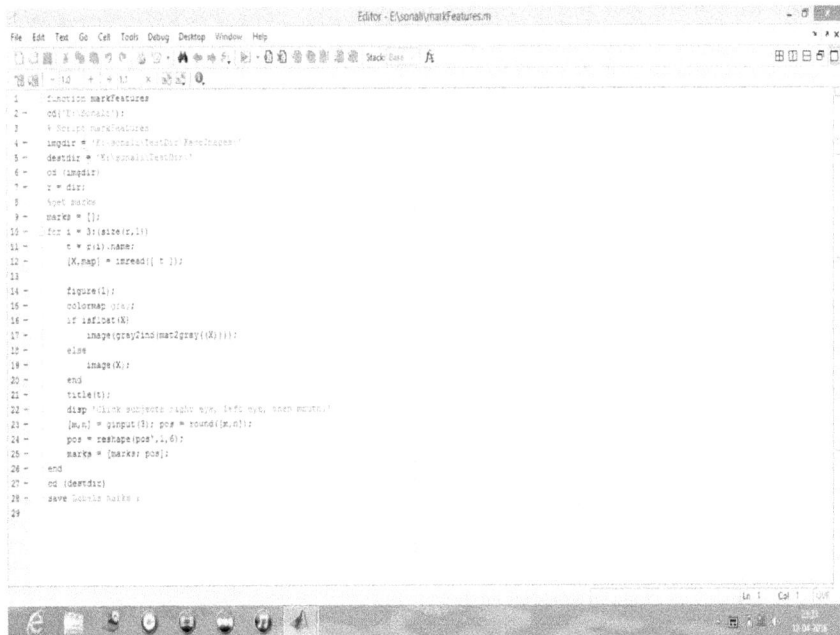

Figure 2.12 To mark facial features.

Figure 2.13 Step 1: marking facial features (left eye, right eye and mouth).

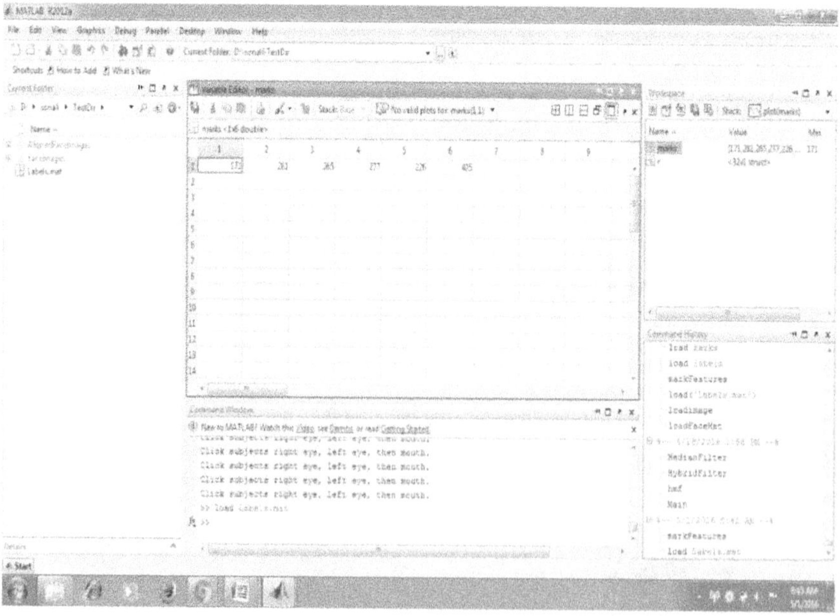

Figure 2.14 Step 2: Matrix form of marked facial features.

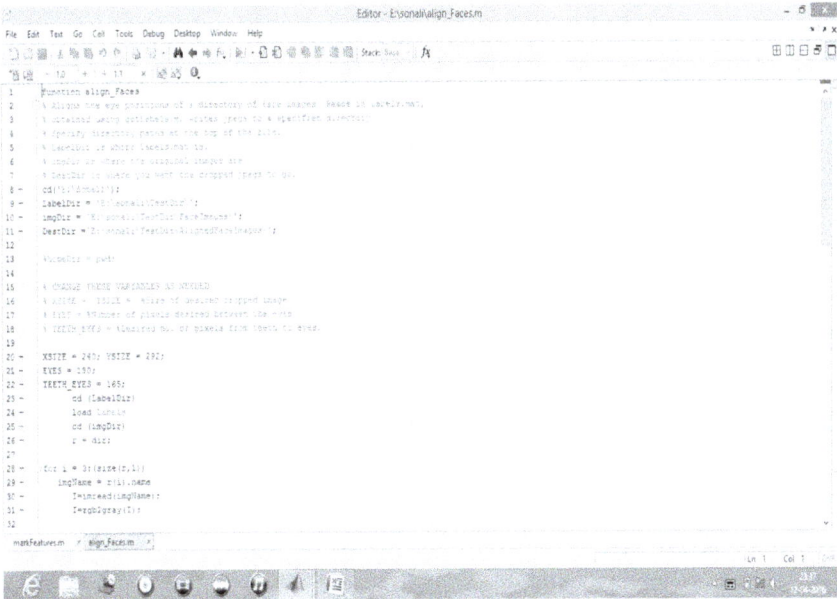

Figure 2.15 To align the faces.

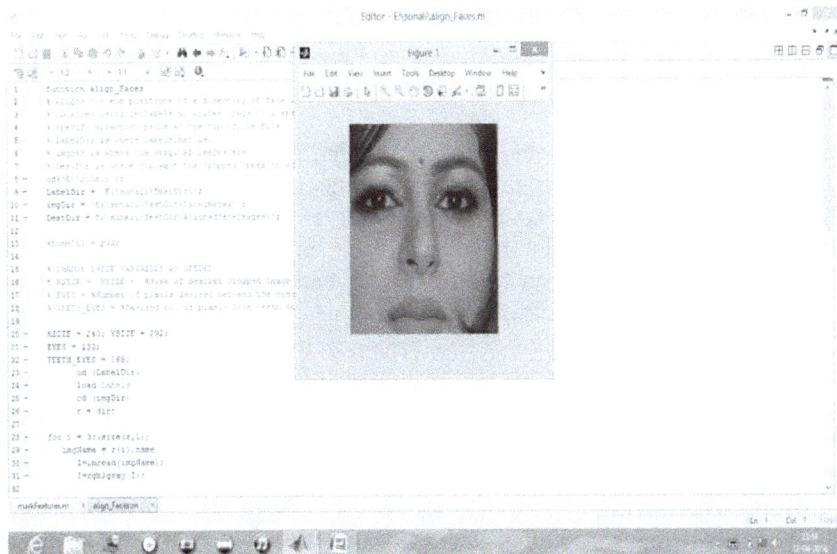

Figure 2.16 Step 3: Format conversion from JPG to pgm for alignment of faces.

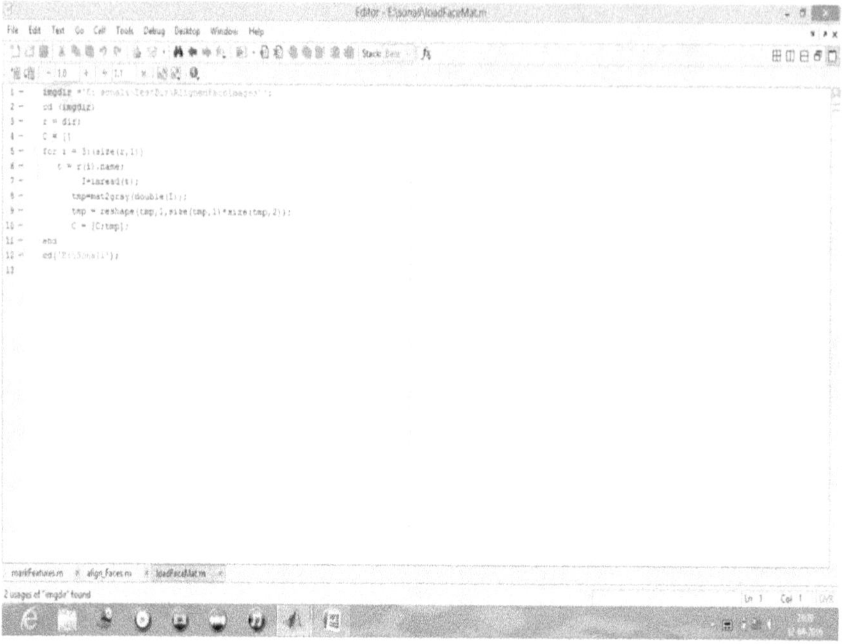

Figure 2.17 Step 4: Matrix formation for all the images.

and loaded into MATLAB tool, it shows that this format is not compatible and that is why conversion is required in this step.

Load Faces- This step loads all the aligned faces into a matrix named as C in which there are 41 rows and 70613 columns. The aim of this function is reshaping, it means to load all faces in the form of a matrix which is shown in Figure 2.17.

Main Function: This is the main step which gives output as how many images are matched with training database means in terms of percentage it is shown that if 41 images are taken for mark features, then from 41 images that is called as training database how many images are there in a matrix created by load faces. Its benefit is to check whether the training database is complete or not before going to next step, i.e., testing of an image. This Figures 2.18, 2.19 and 2.20 show, respectively, the output of the main function which gives pc variable. If the value of pc is 100%, it shows that training database is complete means accuracy of this model is 100%.

Train: Next step is training of the network by taking a feedforward neural network in order to check the performance for the proposed system in MATLAB tool. This step extracts ICA with PCA features from the matrix C of the load faces module as depicts in Figures 2.21, 2.22 and 2.23. In training, input values are stored as input.mat file, output values are stored as targets.mat file, network type is feedforward distributed time delay as it

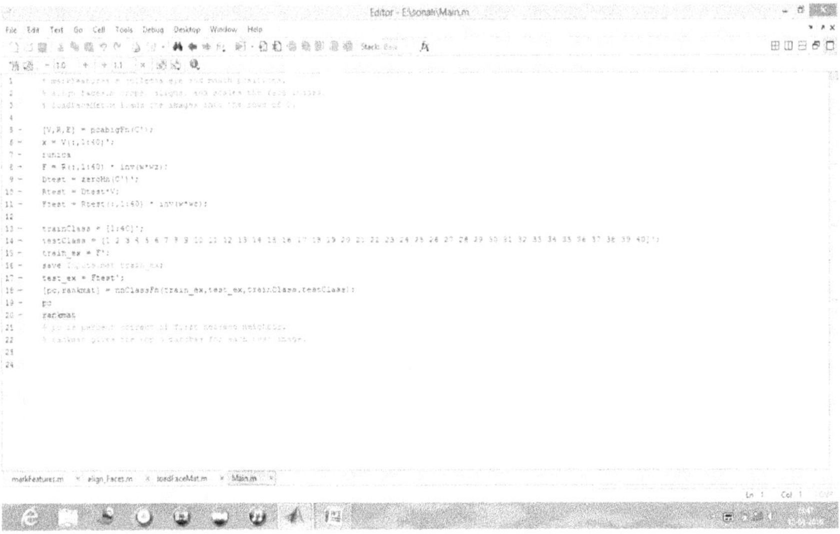

Figure 2.18 To run the main function for verification.

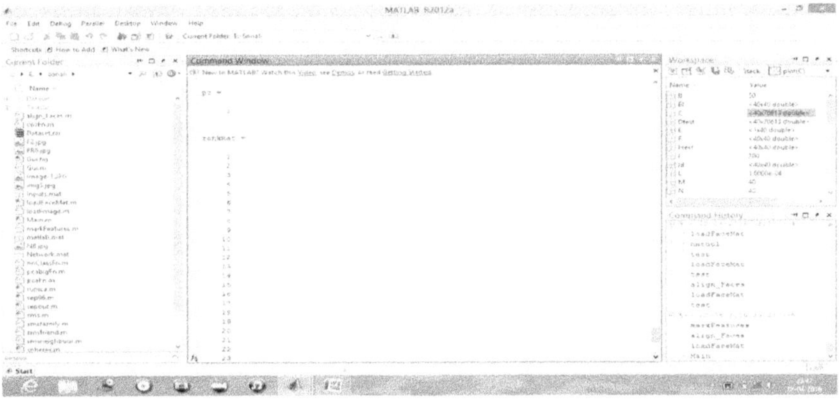

Figure 2.19 Output in form of pc and Rankmat.

is providing maximum performance when compared to others such as back propagation network and hopfield network and training function is trainlm which is a by default function.

Test: This is a step for recognizing a face image that it is known or unknown category and sends text messages to all the person belongs to one's category. For example, in Figure 2.24, if a face is recognized as a neighbor, then the message is sent to all the members of family and neighbor category. In proposed work, there are four categories: family, friend, neighbor and

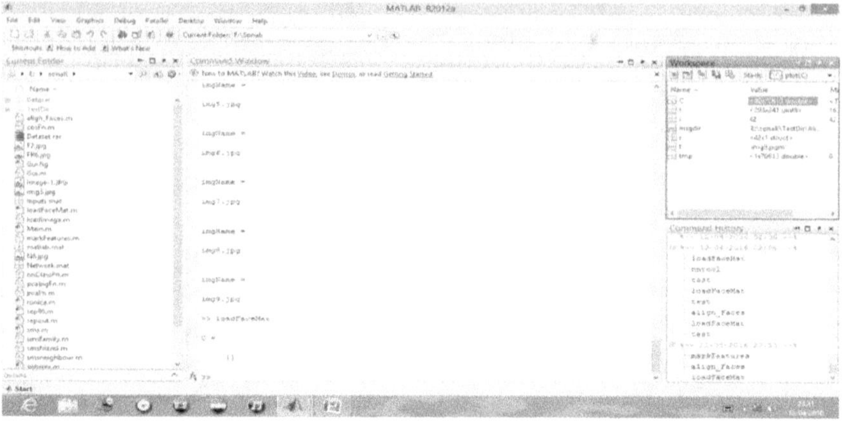

Figure 2.20 Step 5: Database formation verification.

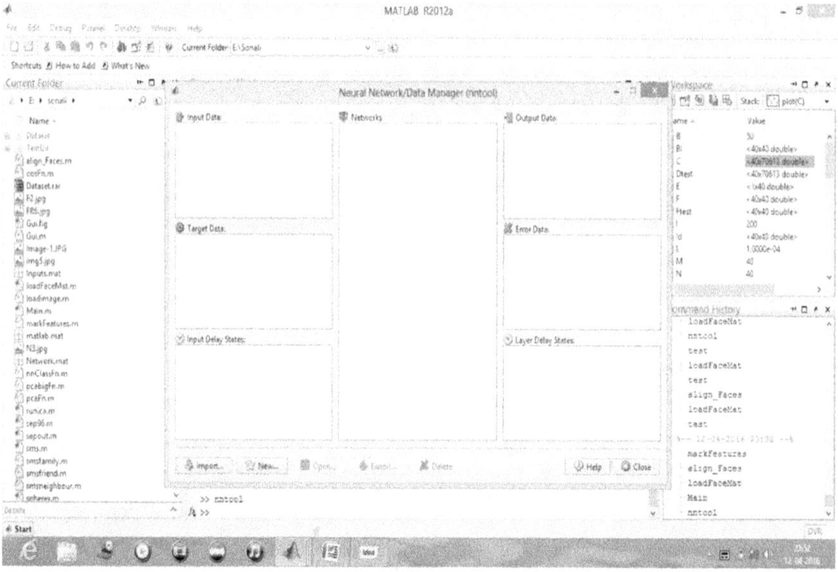

Figure 2.21 To create neural network by nntool in MATLAB®.

unknown. Table 2.1 shows the values that who will be the receiving party in case of face recognition. If the face is recognized as family category, in that case, message will be sent to all family members and neighbors. In the second case, if the face is recognized as belonging to friend category, then message will be sent to all members of friend and family category, it is clear that in

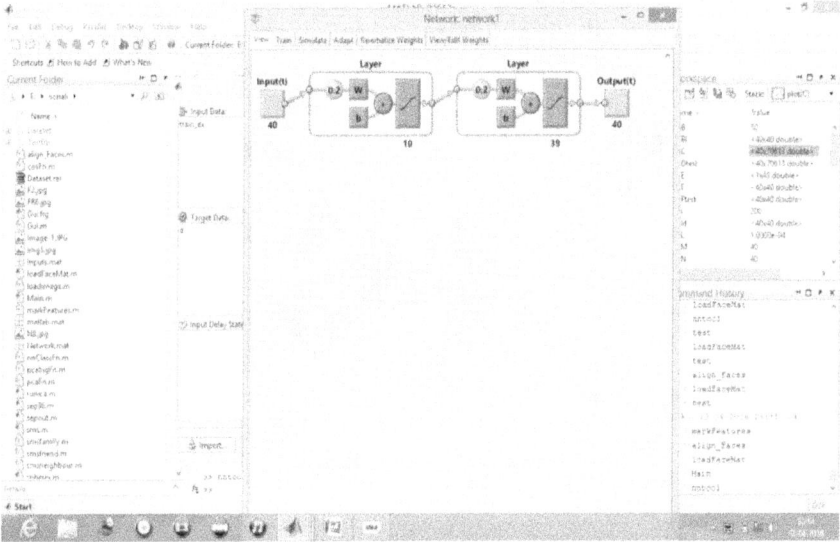

Figure 2.22 Step 6: Image verification using feedforward neural network.

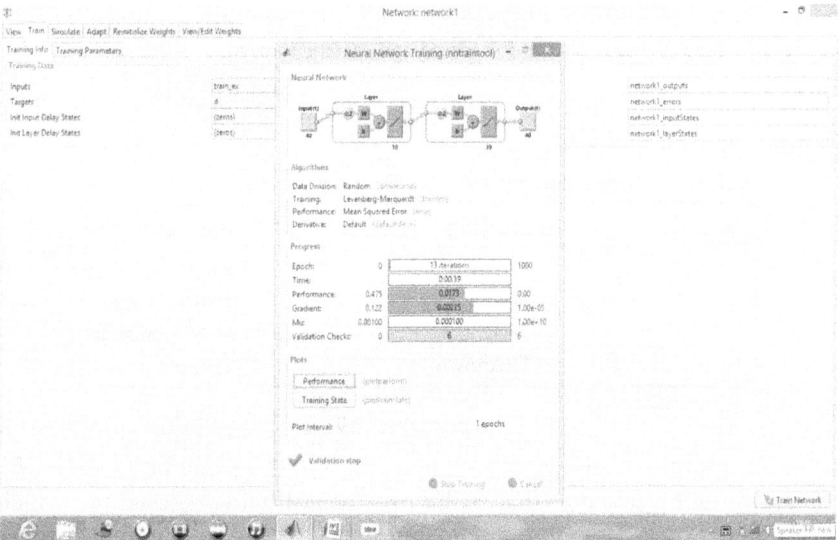

Figure 2.23 Neural network processing complete.

every case the text message will be sent to family members as we are providing security for an individual who is living alone at home. So this step is necessary.

Performance Graph: This step finds the performance of neural network for the proposed work by focusing on objectives of getting least mean

Figure 2.24 Step 7: Result shown for tested and matched image.

Table 2.1 Category corresponding to receiving parties

S. No	Category	Message Parties
1	Family	Family + neighbors
2	Friend	Family + Friend
3	Neighbor	Family + neighbors
4	Unknown	Family

square error by selecting that value of a neural network classifier which provides the train data set, test data and some types of validation below the line of error as shown in Figure 2.25. The concept of epoch and iteration is used in this work by providing low mean square error. After running nntool in MATLAB workspace, load input data named as inputs.mat, output data named as targets.mat and establish a network to train the data and then run this tool to evaluate the performance.

After getting the desired output, the last step is to send messages to re-spective category members on their mobile phones using 3G Dongle and

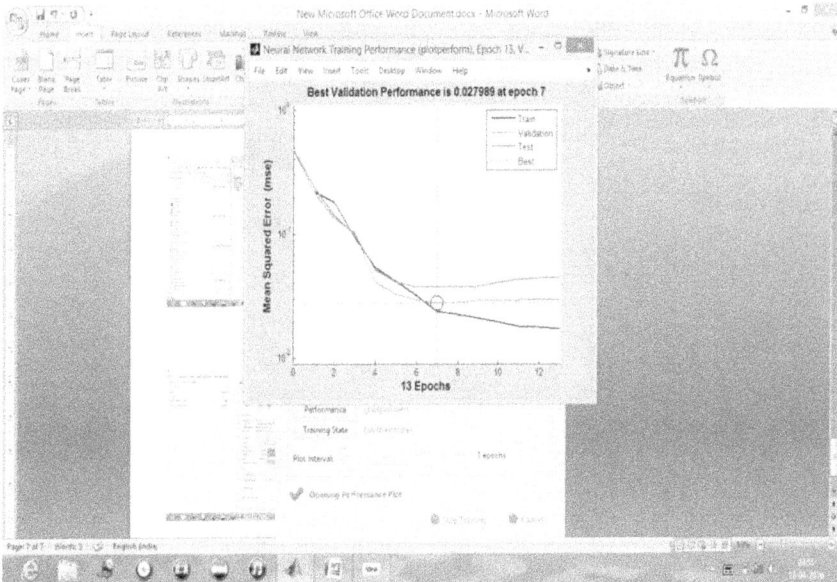

Figure 2.25 Performance graph for image verification and identification

MATLAB tool. The sample images stored in a database is used for experimentation which comprises of 41 images and a database of sample images is shown in Figure 2.26.

2.5 RESULTS

Previously, a number of methods such as PCA, ICA and LDA have been implemented in databases to analyze the efficiency of the system. Table 2.2 shows the comparison of proposed work with individual PCA, Table 2.3 depicts the relation of the proposed work with individual ICA and lastly, Table 2.4 depicts the comparison values of proposed work with individual algorithms. It is concluded that the present hybrid method provides better results than the previous methods.

Figure 2.27 depicts the performance graph for individual PCA with respect to recognition rate, Figure 2.28 describes the performance graph for individual ICA with respect to recognition rate and Figure 2.29 shows the analysis of proposed hybrid PICA algorithm with individual PCA and individual ICA algorithms. In this graph, a number of images are taken as a parameter on x-axis and percentage of recognition rate is taken on y-axis. This graph shows that in the case of the proposed hybrid PICA algorithm, recognition rate is 99.5%.

Figure 2.26 Sample images stored in database (family, friends and neighbor).

Table 2.2 Individual PCA performance in terms of recognition tate

S. No.	Data set	Recognition rate
I	I(10 images)	98%
2	II(20 images)	98%
3	III(30 images)	98%
4	IV(40 images)	97%
5	V(50 images)	93%

Table 2.3 Individual ICA performance in terms of recognition rate

S. No.	Data set	Recognition rate
I	I(10 images)	92%
2	II(20 images)	92%
3	III(30 images)	93%
4	IV(40 images)	98%
5	V(50 images)	98%

Table 2.4 Comparison of PCA, ICA and proposed hybrid PICA algorithm

S. No.	Data set	Recognition rate PCA+NN	Recognition rate PCA+NN	Recognition rate hybrid PICA+NN
I	I(10 images)	98%	92%	99%
2	II(20 images)	98%	92%	99%
3	III(30 images)	98%	93%	99%
4	IV(40 images)	97%	98%	99.5%
5	V(50 images)	93%	98%	99.5%

Desired results are obtained by using a 3G dongle with MATLAB tool in such a way that text messages will be received on cell phones from the number which is registered with the dongle. Messages can be sent to a number of people at the same time by inserting their numbers in code of MATLAB [59]. The number of commands is used to produce operations for

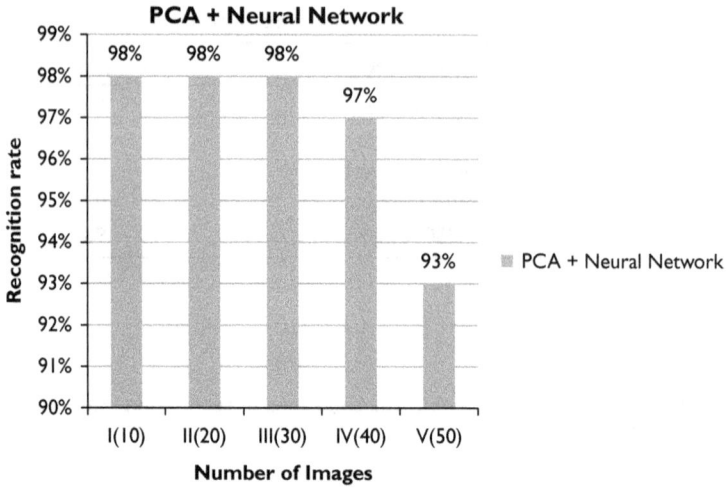

Figure 2.27 Performance graph of individual PCA in terms of recognition rate.

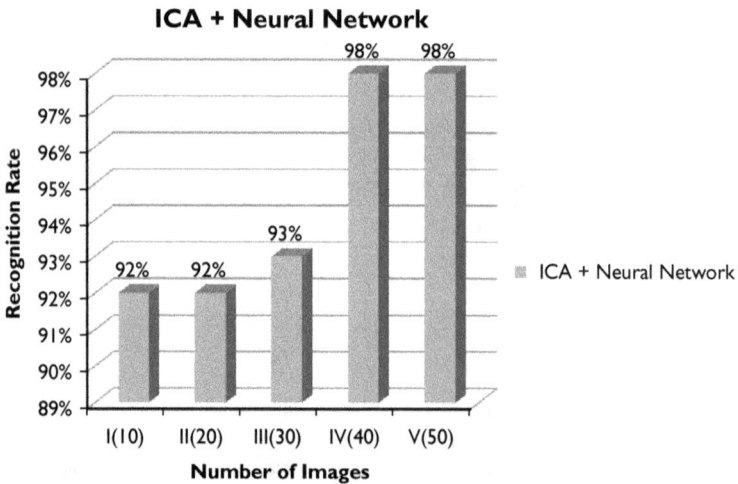

Figure 2.28 Performance graph of individual ICA with respect to recognition rate.

receiving the message. Figure 2.30 shows the screenshot of message on phone received by using MATLAB tools and a 3G dongle.

2.6 SUMMARY

A new technique named PICA has been proposed for human identification using fusion of two existing face recognition algorithms named as PICA,

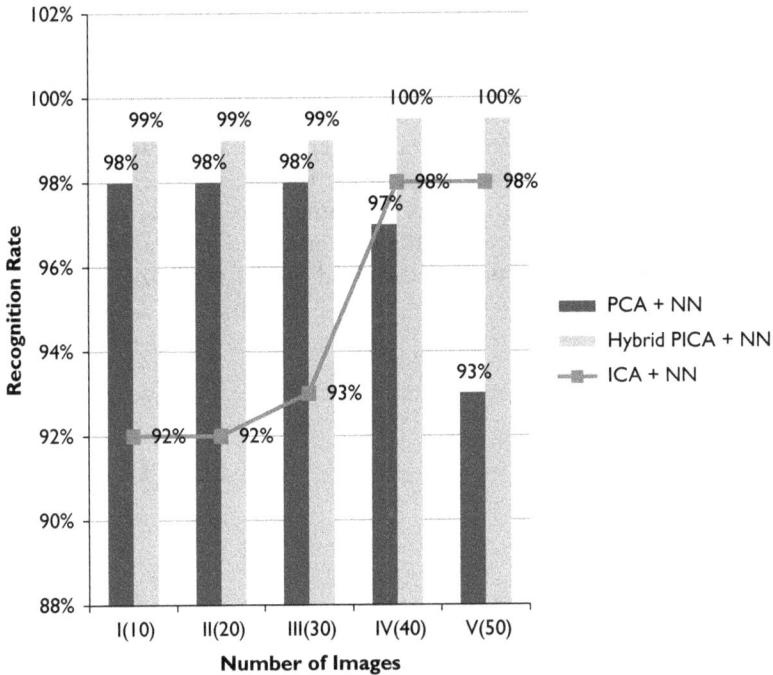

Figure 2.29 Analysis of proposed hybrid PICA algorithm and previous algorithms in terms of recognition rate.

Figure 2.30 Screen shot of text message being displayed on cell phone.

which can substantially improve the rate of recognition as compared to the single recognition algorithm. The proposed system uses a hybrid combination of PCA with Independent Component Analysis (ICA) for extraction of facial features and Back Propagation Feedforward Neural Network is used for recognizing a face. The experimental results show that the hybrid method takes the features of PCA and ICA in feature extraction and gives an excellent performance for person's face recognition. Extensive experiments on data sets are carried out to calculate the performance of various features. In the future, security measures may be taken into consideration while taking and storing the images of the person, face recognition can be applied to iPhone security to unlock the phone. Face unlock is a feature that allows unlocking the phone using the face print and integrated sensors may also be embedded in the proposed hybrid model for providing safe and convenient environment.

Chapter 3

Hybrid **MFRASTA** voice recognition technology for individual's security at home

3.1 INTRODUCTION TO VOICE RECOGNITION

Voice is a name given to the sound that carries the contents of language. People learn a number of skills during childhood, and this process continues throughout the life. It seems very natural to us as voice phenomena's complexity is not realized. There are two biological organs named as: human vocal tract and articulators having properties that are affected by gender to emotional state and vocalizations are varied in terms of volume, speed, accent, pitch, pronunciation, etc. The total data amount that is generated during the voice production is very large and not all of them contain useful information. So, less data are required to represent characteristics of voice and the person who has spoken it. Voice recognition is a method used to recognize a word automatically which is spoken by any speaker. Voice recognition is also identified as speech recognition. It is a software program that has the ability to decode a person's voice. The attractive feature of voice recognition is without using any mouse, keyboard or any type of button, it operated devices and can perform any type of commands. The concept of voice recognition was originated by the way human communicate to each other. No two individuals have similar voice and the difference lies in the construction of articulator organs. Length of vocal tract, shape of mouth, features of vocal cord make voice of every user unique.

The research and development process of voice recognition technology started in the early 1970s with the help of researchers at IBM Corporation and Carnegie Mellon University and first automatic speech recognition was used in 1952. Humans can also communicate with machines through voice but that is a very slow process. So, automatic voice recognition comes there that can be operated by human commands or instructions. By using this technique, the person's voice can be verified in order to find his identity which will be used in providing access to applications like database services, voice mail, telecommunications (it means in cell phones where a person can dial a number by just speaking the name of the person he wants to call), etc. Voice recognition problem is solved by understanding its complexity. There

DOI: 10.1201/9781003120933-3

are number of areas where automatic voice recognition has been applied: healthcare, personal computing, industries, etc. This concept is used in automated phone systems, which provide assistance for directing the call to the caller with the correct department. For example, if a command is there that "press number 3 for support" and a person says 3, it means concept of voice recognition is used here. Another example is of Google voice that allows a person to ask and search questions from laptop, ipad or tablet and digital assistant is also a part of voice recognition, which helps the users to interact with digital assistants that help in answering the questions by Siri, Amazon Echo, etc. For proper working of voice recognition, a computer is required with a soundcard, headset or microphone. There are four parameters that are considered for the operation of the voice recognition system: it must have encoding form and number of utterances that it will recognize, at the time of recognition there must be a type of a pattern matching algorithm which compares the representation of the input utterance with all the utterances in the vocabulary, compares input utterance with vocabulary by using any algorithm and last is an interface which operates the recognition system. There exist a number of other techniques for biometric authentication named as face, signature, iris recognition, etc. But voice recognition technique is selected on the basis of a criterion: accessibility, distinctiveness, robustness, etc. Distinctiveness is used to measure the differences in the patterns among the individuals, robustness is no repeatable pattern and cannot be subjected to large changes and accessibility defines the easiness for presentation of data to sensor device [60]. Recognition through a voice is very economical as the equipments for collecting speech samples are cheaply and easily available. Figure 3.1 shows the general diagram of speech recognition system where input speech goes to acoustic model block in which a transformation takes place from speech to statistical representation of vectors. Then, a word is searched with number of words stored in the database and provides output of recognized utterance.

This voice recognition system mainly consists of two steps:

- Feature extraction
- Feature matching

Extraction of features is the main step used as a front end. It acts like a heart of the system. It extracts little quantity of data from the words which is then used later to express each person. Features are high level representations as

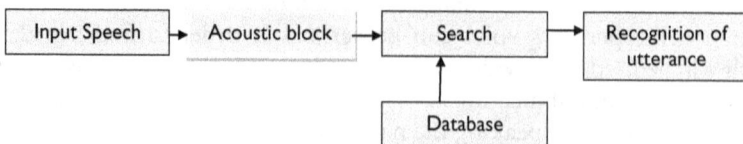

Figure 3.1 General diagram of speech recognition system.

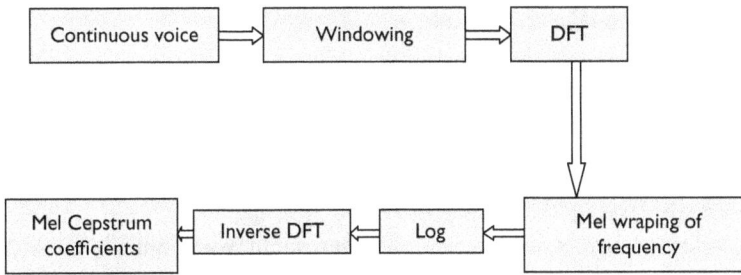

Figure 3.2 Steps for feature extraction.

compared to raw data representations. It compresses the magnitude of words without causing any type of harm to the speech signal power. Below is the feature extraction diagram in Figure 3.2, which shows the basic steps require for extracting the features:

Matching of features is used to determine unknown identity by comparing the features that have been extracted with a complete set of input samples. It means this system works on two phases: training phase and testing phase.

In training phase, it corresponds to the procedure of learning from labeled data and each individual will provide samples of their own voice in a way to train the system for another processing as mentioned in Figure 3.3.

In the testing phase, when a person's voice is tested, input voice is matched with all stored samples and on the basis of that, a decision is taken. This process is shown in Figure 3.4.

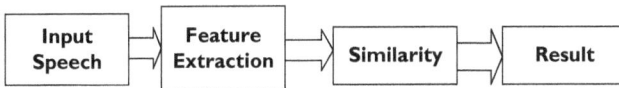

Figure 3.3 Training phase for extraction of voice features.

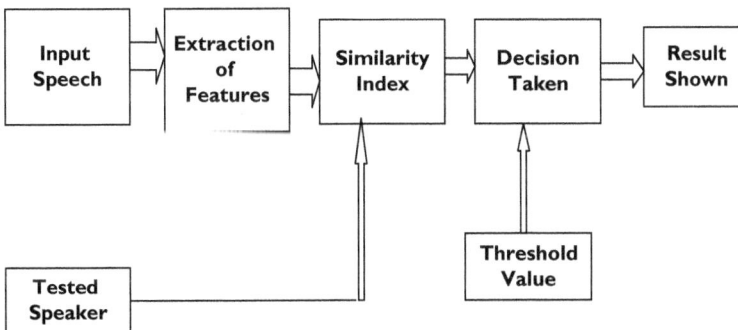

Figure 3.4 Testing phase for feature matching.

Basic terms used in this system are:

- **Utterance:** It is defined as the word spoken that represents some meaningful information to the computer system. It may be an individual word or set of words and it can be a sentence. They are then sent to the speech engine for further processing.
- **Vocabulary:** It is a group of words that are recognized by voice recognition system. It is obvious that small vocabularies are easy to recognize while large vocabularies are difficult to recognize.
- **Accuracy:** It is defined as the ability of a recognizer that how correctly it recognizes the words. It is used to identify an utterance.

Voice recognition system can be classified in a way that how many spoken words are recognized:

- **Isolated word:** It needs each and every word to have silence on adjacent parts of the window. At a time, it works on an individual word: it means a pause is required between the words. It is the simplest form of voice recognition as the end the points are easier to find.
- **Connected words:** It is same as isolated word, but it allows separate words to run together with a minimum time gap.
- **Continuous speech:** It is not separated by pauses; it is based on the voice where words are combined together. In this, the person speaks in a natural way, and it is difficult to recognize because it needs methods to define boundaries of words. It can be used in voice repertory dialler where eyes free, hand-free dialling is possible.

There are a number of techniques that are used to analyze a voice waveform:

- **Waveform:** Voice signal is a sequence of pressure variations in the part between listener and the sound source. In this graph, the horizontal axis is the axis for time parameter from left to right, and the curve indicates the variation of pressure parameter in the signal.
- **Pitch:** Voice is a process consisting of two parts: sound source and filtering. It directs to take the fundamental frequency of the sound when the ultimate voice utterance is analyzed.
- **Spectrum:** It gives a representation of a frequency distribution with the amplitude of a moment in time.
- **Cepstrum:** This technique is used for speech analysis, it is based on the spectral view of the signal.

There are a lot of advancements in voice recognition technology. Previous methods were well defined under all conditions, but they were not providing good recognition rate results if the single feature extraction technique was used in voice recognition, so there is a need for improvement.

The basic aim of the proposed research work is to generate a system that will provide high recognition rate with high accuracy and efficiency in order to provide security for an individual who is living alone at home. To achieve this objective, a hybrid MFRASTA algorithm for feature extraction is implemented that gives useful results than existing ones. The hybrid algorithm is implemented using MATLAB® tool.

3.2 CHALLENGES OF VOICE RECOGNITION SYSTEM

The voice recognition system is classified by speaker-dependent and speaker-independent systems: speaker-dependant system recognizes the words spoken by a preregistered user by comparing them with stored samples of his/her voice in database, and speaker-independent system recognizes the words without prior experience with that particular person. There are two major factors that make a challenging situation for voice recognition system: loud environment and reach factor. There exist a number of parameters that affect the reliability and performance of voice recognition system:

- A speaker voice variation according to emotions, gender, speeds of speaking, accent or pronunciation
- Environment variation according to noise
- Transmission channel variation
- Accuracy
- User responsiveness
- Performance
- Reliability
- Fault tolerance

3.3 VOICE RECOGNITION ALGORITHMS

Voice recognition is a technique of automatically verifying the speaker by taking information of an individual present in voice waveform. Two types of voice frames are obtained: spectral and prosodic. Spectral features can be determined by cepstral analysis in which FFT is calculated from input samples. Then, power spectrum is calculated and log value is applied to it. Prosodic features are not having any direct correspondence to write characters of a sentence. There are a number of algorithms used for recognition of the voice. Below is the explanation of all the algorithms with their features and drawbacks.

3.3.1 Mel frequency cepstral coefficient (MFCC)

It is a popular technique developed by Davis and Mermelstein in 1980s. As we know, the important step in automatic recognition of voice is to obtain the features, i.e., to identify the elements of the audio signal which is good for identifying the linguistic content and discarding other stuff which carries information such as background noise and emotions [61]. MFCC is based on domain of frequency using Mel scale, which is further based on human ear scale. It is an audio-based feature extraction technique used for extracting the parameters from voice similar to the ones that are used by humans for hearing speech and at the same time, deemphasizes all other information [62]. The block diagram of MFCC is depicted in Figure 3.5.

Preprocessing: First, voice signal is provided as an input, then it will preemphasized, in which high frequencies are artificially amplified as shown in Figure 3.6. In this, framing means to multiply the signal with a number of rectangular windows in sliding state, but the difficulty with rectangular window is of high power involved in side lobes that's why it may give originate to spectral leakage. In order to prevent this problem, the concept of hamming window is used. Filter coefficients of hamming window having length n are computed as

$$W(n) = .54 - .46 \, Cos\left(2 \prod n/N - 1\right)$$

N = Total number of samples
n = Current sample

Mel frequency cepstral coefficient: They are calculated using triangular-shaped bank filters with central frequency of filter spaced less than 1000 Hz linearly and above 1000 Hz logarithmically. Each filter is computed by its bandwidth using central frequency of two filters adjacent to each other, and it is dependent on the frequency band of filter bank along with a number of filters that are chosen for the design.

Silence detection: An important step that is necessary in front-end voice recognition system. When a word is spoken by a person, at that time the system had no control, that's why this step is must to determine at what time the person has actually started spoken the word and immediately converts the frame of that word for an instant.

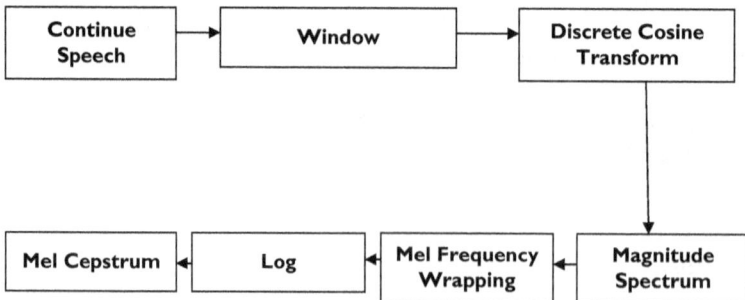

Figure 3.5 Block diagram of MFCC technique.

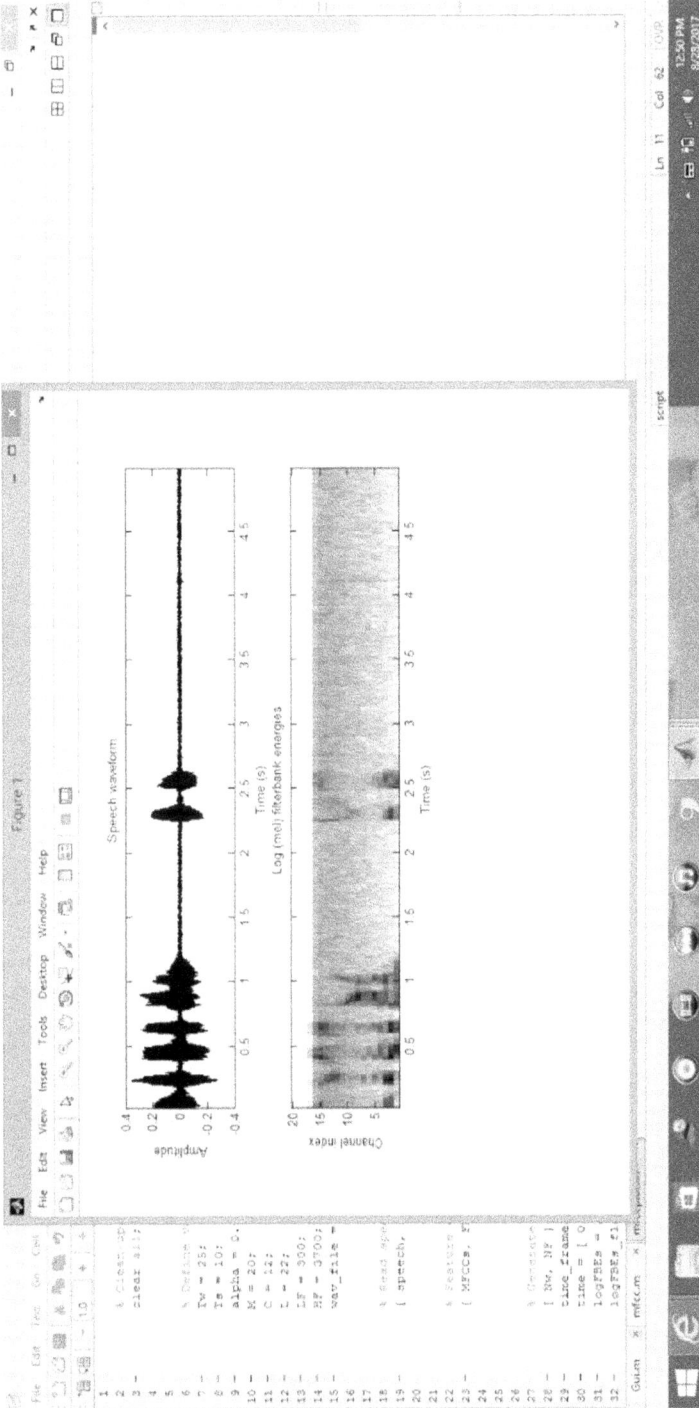

Figure 3.6 MFCC preprocessing phase.

Windowing: This is an important step in which each individual frame is windowed in order to minimize the discontinuities of the signal at the beginning of the frame and at the end of the frame. In this, the criteria used to minimize the discontinuity or spectral distortion is using the window in order to tape the signal by a zero value at the beginning and end of the frame. Window in MFCC is described as

$$W(n), 0 < =n < = N - 1$$

where N = number of samples in each frame

$$Y1(n) = x1(n) \, w(n)$$

where $0 < = x < = N - 1$

FFT: It converts each frame of a number of samples represented by N from time domain to another domain named as frequency domain. It is a fast algorithm used to implement Discrete Fourier Transform, which is defined on set of N samples $\{X_n\}$ as follows:

$$X_n = \sum_{k=0}^{N-1} X_k e^{-2\prod jkn/N}$$

where n = 0, 1, 2, 3 ... N - 1
j denotes the imaginary part

Mel frequency warping: Linear scale is not followed by voice signals containing frequency content of sounds. So, for each tone having actual frequency f is measured in Hertz, another pitch is measured on a better scale called Mel scale. It is a part of frequency spacing below 1000 Hz linearly and spacing above 1000 Hz logarithmically. A formula is used to compute the Mel coefficients (Mels) for a given frequency represented as f in Hz.

$$Mel(f) = 2595 * \log_{10}(1 + f/700)$$

Cepstrum: This parameter converts log Mel spectrum back into time and the output is called as MFCC. Cepstral representation of speech spectrum provides a good representation of local spectral properties of signal for giving frame analysis. To calculate MFCC's C_n as

$$C_n = \sum_{k=1}^{k} ((\log S_k) \, Cos \, [n \, (k - 1/2) \prod /k]$$

where n = 1, 2, 3 ... k.

Advantages:
- It is used for speech processing tasks.
- Its recognition accuracy and performance rate are high.
- It has low complexity.

Disadvantages:
- This technique includes little noise robustness.
- As it takes into account only the power spectrum and ignores the phase spectrum, hence it provides limited representation.
- It fails to recognize same word uttered with different energy.

3.3.2 Perceptual linear prediction (PLP)

PLP is similar to linear prediction code that is based on the short-term spectrum of speech, but this technique is included with some better methods using several psychophysically based transformations [63]. In this technique, three concepts are used to derive an estimate of the auditory spectrum.

- Critical Band Spectral Resolution
- Equal Loudness Curve
- Intensity Loudness Power Law

It rejects the irrelevant information of speech and thus improves the voice recognition rate.

A frequency warping [63] is applied to Bark Scale filter bank. In this technique, first step is to convert frequency into Bark Scale which is a better representation of human hearing resolution in frequency bark and this representation, as shown in Figure 3.7. The frequency corresponding to an audio frequency is

$$\Omega(w) = 6 \ln [w/(1200 \prod) + [(w/(1200 \prod))^2 + 1]^{0.5}]$$

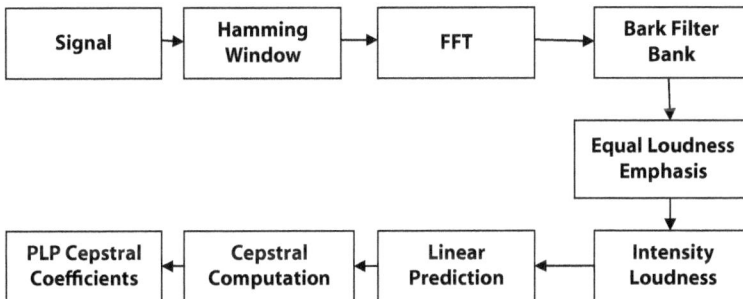

Figure 3.7 Block diagram of PLP technique.

Bark Filter Bank consists of integration of three factors into one filter bank: frequency warping, smoothing and sampling. An equal loudness pre-emphasis weights the filter bank outputs to simulate the sensitivity of hearing. Equalized values are transformed according to a power Law of Stevens by raising each to the power of 0.33. Resulting auditory warped line spectrum is further processed by linear prediction. Applying linear prediction to auditory warped line spectrum means to compute predictor coefficients of a signal. Finally, cepstral coefficients are obtained from predictor coefficients by a recursion that is equal to log of the model spectrum followed by inverse Fourier transform.

Advantages:
- It reduces the discrepancy between voiced and unvoiced speech.
- Its resultant feature vector is low dimensional.
- It is based on the short-term spectrum of speech signals [63].
- The coefficients of PLP are mainly used as they provide correct value of high-energy regions of the voice spectrum.
- LPC approximates the voice spectrum in equal manner for all frequencies, and this presentation is required to known principles of human hearing.

Disadvantages:
- Its resultant feature vector is dependent on whole spectral balance of formant amplitudes.
- Its spectral balance is easily altered by the communication channel, noise and equipment used.

3.3.3 Linear prediction code (LPC) [63]

LPC is the most important feature for processing of voice signals that takes out the voice parameters such as pitch formants and spectra. This technique is used for designing user's voice production. It predicts the future features based on previous features. It is desirable to compress the signal for efficient transmission and storage. The digital signal is compressed before transmission for efficient utilization of channels in wireless media. For medium or low bit rate coder, it is widely used. While we pass the voice signal from the voice analysis filter in order to remove the redundancy in signal, residual error is generated as an output. It can be quantized by a smaller number of bits as compared to the original signal. So now, instead of transferring the entire signal, transfer this residual error and speech parameters to generate original signal. A parametric model is computed that is based on least MSE theory, this technique is known as linear prediction. Figures 3.8 and 3.9 represent the block diagram and steps included in LPC technique.

   i. Preemphasis: First step is speech analysis that is performed by passing it with the help of a filter, and the goal is to flat the signal in spectral

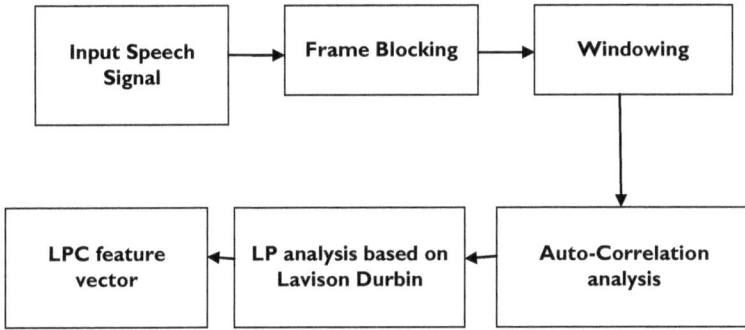

Figure 3.8 Block diagram of LPC technique.

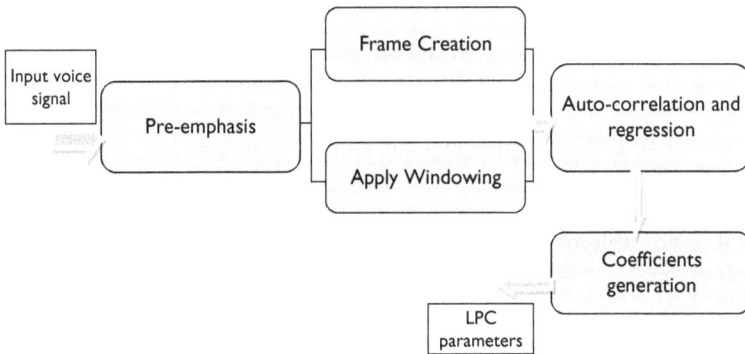

Figure 3.9 Block diagram of LPC technique.

form with less precision effect. The value of filter coefficient must be within 0.9 and 1. [64].

ii. Framing creation: After first step, the developed voice is divided into a number of frames that consist of N samples, and the range of each sample is 20 to 40 seconds. A must requirement is a standard overlap of 10 ms within two nearby frames in order to ensure the stationary period between frames.

iii. Apply windowing: This step takes resulting frames, and they are multiplied with a hamming window. The main purpose is to minimize the edge effect.

iv. LPC computation: This is the last and final step in which the technique of autocorrelation is applied on the frames of windowing speech sample.

It analyzes the speech signals by estimating the formants, removing their effects from the speech signal and estimating the intensity and frequency of remaining buzz. The process of removing the formants called inverse filtering and remaining original is called residue.

Types of LPC [64]:
There are a number of LPC techniques which are named as follows:

- Coded-Excited LPC (CELP)
- Pitch Excitation LPC
- Residual Excitation LPC
- Voice Excitation LPC
- Multiple Excitation LPC (MPLPC)

Advantages:
- It is a static technique.
- It is reliable, accurate and robust technique for providing parameters of speech.
- Computation speed is good and provides with accurate parameters of speech.
- It is used for encoding of speech at low bit rate.
- It is one of the powerful ways of signal analysis.
- It minimizes the sum of squared differences between the real speech signal and calculated signal for a given time period.

Disadvantages:
- It is not able to distinguish the words with similar vowel sound.
- It generates residual error as output, resulting in poor speech quality.

3.3.4 RelAtive SpecTrA-Perceptual Linear Prediction (RASTA-PLP)

This method is used to enhance the speech when recorded in a noisy environment. In this, RASTA technique is merged to the PLP technique and forms a combination named as: RASTA-PLP, this combination increases the robustness includes in individual PLP method. The fact used in this method is that the temporary properties of environment surrounded by persons are not same with the properties of signals of voices [65]. When combined with PLP, it gives better performance. Proposed by Hynek Hermansky, RASTA-PLP is a method of encapsulating the spectra for difference minimization between speakers so that useful information is preserved. RASTA is a useful technique applied to each sub band of frequency by using band pass filter and finally person got smooth signals irrespective of noise factor [65].

Figure 3.10 displays the most useful processes that are defined in this technique. This technique calculates the critical band power spectrum as in PLP, transforming spectral amplitude through a compressing static non-linear transformation, filtering the time trajectory of each transformed spectral component by band pass filter using an equation, i.e.,

$$H(z) = (0.1)(2 + z^{-1} - z^{-3} - 2z^{-4}/z^{-4}(1 - .98z^{-1}))$$

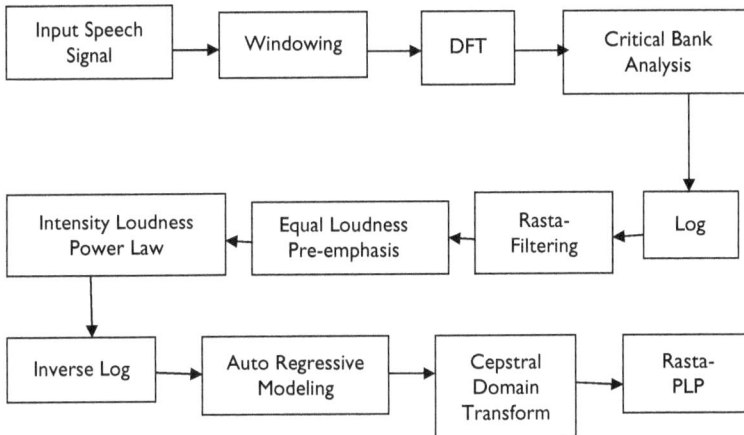

Figure 3.10 Block diagram of RASTA-PLP technique.

Advantages:
- It is robust.

Disadvantages:
- It gives poor performance in the clean speech environment.

3.3.5 Zero crossing peak amplitudes (ZCPA)

This technique was developed by Chok-ki Chan. It is based upon a speaker-dependant system and human sound system [65]. ZCPA uses intervals with crossing zero to express the information about signals and provides value of amplitude to express intensity information. With this technique, value of frequency and amplitude is combined to form the complete output of feature vectors. This technique classifies a set of words into two groups and makes the process of recognition easier as presented in Figure 3.11.

3.3.6 Dynamic time warping (DTW) [66]

In time arrangement analysis, this is a calculation for estimating similarity between two temporal successions which may change in speed as well as in time. For example, by using this algorithm, walking pattern' similarity could be easily detected, regardless of whether one individual was walking quicker than the other and in second case, if there were increasing and decreasing of speeds throughout an observation. This technique has been applied to transient sequences of video, sound and graphics data — in

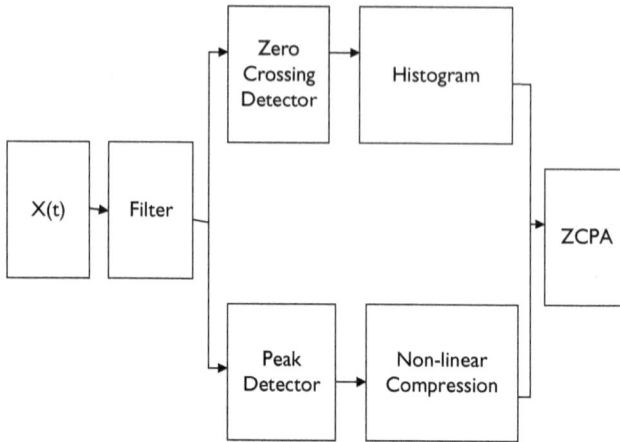

Figure 3.11 Detailed diagram of ZCPA technique.

reality, any information that can be transformed into a linear sequence can be easily analyzed with dynamic time arping, as shown in Figure 3.12.

Advantages:

- High recognition speed.
- Less space is required for the reference template.
- Optimal path finder with constraints
- High recognition rate.
- In the case of error, threshold is used.

Disadvantage:

- Difficult task for choosing a appropriate reference.

3.3.7 Wavelet

It is already known that the signal in a speech is nonstationary [66]. For nonstationary signal, Fourier transform isn't suitable for analysis purpose, and the result is it only provides the frequency data for a signal, but it doesn't give the data about time that at what time which frequency is present. So, the windowed short-time fourier transform is used that provides the temporal data related to the frequency content of the signal.

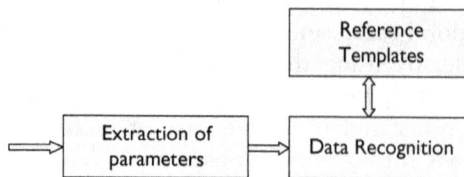

Figure 3.12 Basic diagram of DWT technique.

Applications:
- Signal processing.
- Data compression.
- Speech recognition.
- Computer graphics and multifractal analysis.

3.4 PROPOSED HYBRID MFRASTA TECHNIQUE

Voice signal does not only give information about words or sentences being uttered but also gives the status of the speaker. Voice recognition is an effective way to provide useful information. Feature extraction implements an important role in calculating the effectiveness of the voice recognition system. It is necessary to eliminate redundant information from voice, left with only relevant features. In the proposed work, audio files of known and unknown persons are recorded in.wav format. Two functions named as: voicecompare and voicecompareH are used. The main function used for this work is voicecompareH. In all the previous works, function named as voicecompare is used in which two audio files are compared by using single feature extraction technique, i.e., MFCC but in the proposed work, voicecompareH module is used which is a hybrid combination of MFCC+RASTA-PLP technique. The technique used for comparison of audio files is shown in Figure 3.13 based upon which a decision is taken. This process starts with a person; person will speak a specific password question with its answer. If that question and answer

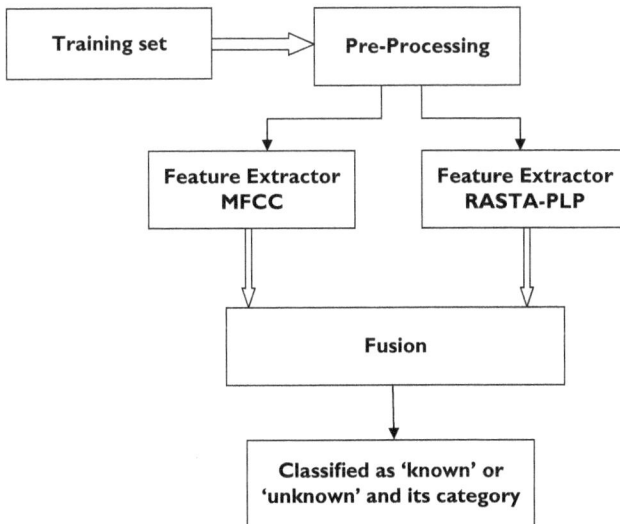

Figure 3.13 Hybrid MFRASTA method for voice recognition.

are similar to the stored value of the database, then only a person is recognized. Finally, the category to which he/she belongs from four categories proposed in a database known as family, friend, neighbor and unknown is identified.

Traditionally, voice recognition system consists of a single technique of feature extraction with classifier, but in this proposed work, there are four steps to be followed:

- Take training set and perform preprocessing by which audio files are recorded using different methods in.wav format.
- Feature extraction step extracts the features that provide specific information from the voice signals. Features have been extracted by using two voice recognition algorithms MFCC and RASTA-PLP.
- Combine all the outputs and classifies input voice that it is known or unknown.
- Send the message to define the category of each input voice. The steps of this hybrid algorithm are represented in the form of a flowchart that is shown in Figure 3.14.

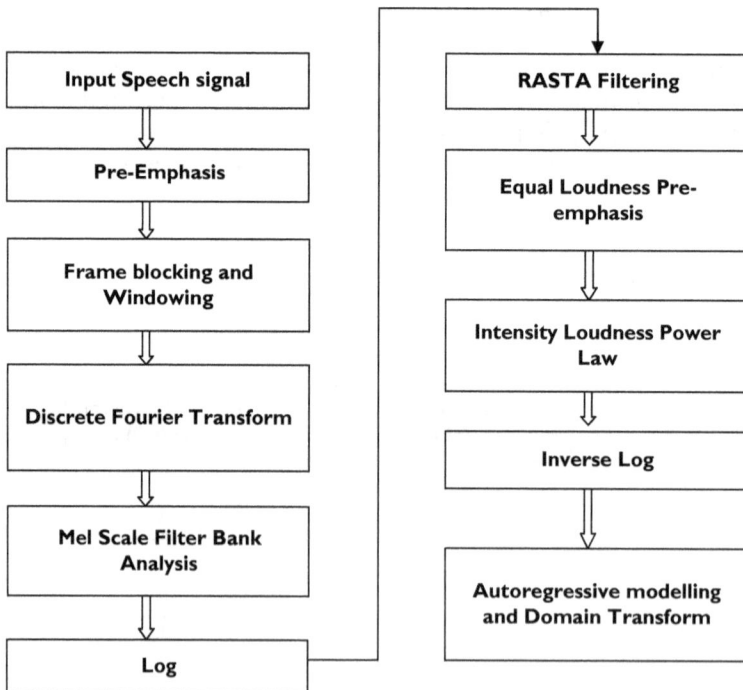

Figure 3.14 Flowchart of Hybrid MFRASTA technique.

3.5 SIMILARITY INDEX

This step will run only in that condition if one of the two biometric technologies (face or voice) fails in recognizing a user who is at the door. In this step, the person's audio file is matched with all the audio files of persons, which are stored in a database on the basis of text-independent voice recognition technique. Text independent means to recognize the voice of a person based on the number of voice functions such as frequency, pitch, speaking rate, etc. irrespective of the content spoken by the person. It means from three modules: face, voice and similarity index, two modules must be satisfied and then only access will be granted to the person. For calculating the similarity index, the value of the threshold is taken and when this step is run, it calculates the value of similarity index. If the value is larger or equal to 75%, it is declared that the person belongs to a known category. But in another case, if the value of similarity index is less than 75%, then a person is supposed to belong to unknown category and in every case, message will be shown on the cell phone of the person who is living alone at home. Decision of opening the door depends upon the individual now.

3.6 IMPLEMENTATION

Integration of Mel Frequency Cepstral Coefficient and Relative Spectra: Perceptual Linear Prediction technique [67] is used for extraction of features and recognition purpose. The steps required to recognize voice and to send messages on the cell phone using a MATLAB tool and a 3G Dongle are as follows:-

1. Voice Recognition: In MATLAB, two processes are used for voice recognition.
   - Voice_Rec_Pre: Its purpose is to record audio files at a frequency of 5000 in.wav format. In this step, preprocessing is performed using two functions: silence detection and amplification from which final file is obtained.
   - VoiceCompare: It is used to compare two wav files. The.wav file to be checked for recognition is correlated with all the audio files saved in the database on the basis of parameters such as frequency range, cepstral coefficients value, etc. If the parameter values are matched with values of any stored wav file, category to which the person belongs out of four considered categories will be displayed as shown in Table 3.1: friend, family, neighbors and unknown. Finally, a message is sent to one or more concerned categories by using MATLAB tools and the Dongle.

Table 3.1 Categories with their corresponding receiving parties

S.No.	Category	Message parties
1	Family	Family + neighbors
2	Friend	Family + friend
3	Neighbor	Family + neighbors
4	Unknown	Family + neighbors

2. In this step, a database consisting of any number of audio files containing voice details of all the persons from all defined categories can be maintained. In the present scenario, 41 audio files belonging to 41 persons considered under three categories, i.e., family, friend and neighbor, and 15 audio files containing voice details of persons belonging to unknown category have been taken into consideration. A sample of audio files is shown in Figure 3.15.

3. **Text or content-based recognition:** The voice of the person at the door is captured and compared with all the stored voices to find out the category to which the person belongs as the database consists of a number of audio files belonging to different categories named as friend, family, neighbors and unknown, as shown in Figure 3.16.

In Figure 3.17, a person at the door belonging to family category shown as f = 1 is obtained as the output. When both techniques face recognition as well as voice recognition will identify a person correctly, a message will be sent to all the family members along with all neighbours who are already registered in the database. In this step, voice is tested on the basis of content which is spoken by the person that means this step is based on text-dependant voice recognition technology.

Nr14.wav Fm1.wav Fm2.wav Fm3.wav Fm4.wav Fm5.wav Fm6.wav Fm7.wav Fm8.wav

Fm9.wav Fm10.wav Fm11.wav Fm12.wav Fm13.wav Fm14.wav Fm15.wav Fm16.wav Frd1.wav

Frd2.wav Frd3.wav Frd4.wav Frd5.wav Frd6.wav Frd7.wav Frd8.wav Frd9.wav Frd10.wav

Frd11.wav Nr1.wav Nr2.wav Nr3.wav Nr4.wav Nr5.wav Nr6.wav Nr7.wav Nr8.wav Nr9.wav

Nr10.wav Nr11.wav Nr12.wav Nr13.wav

Figure 3.15 Sample of audio files stored in database.

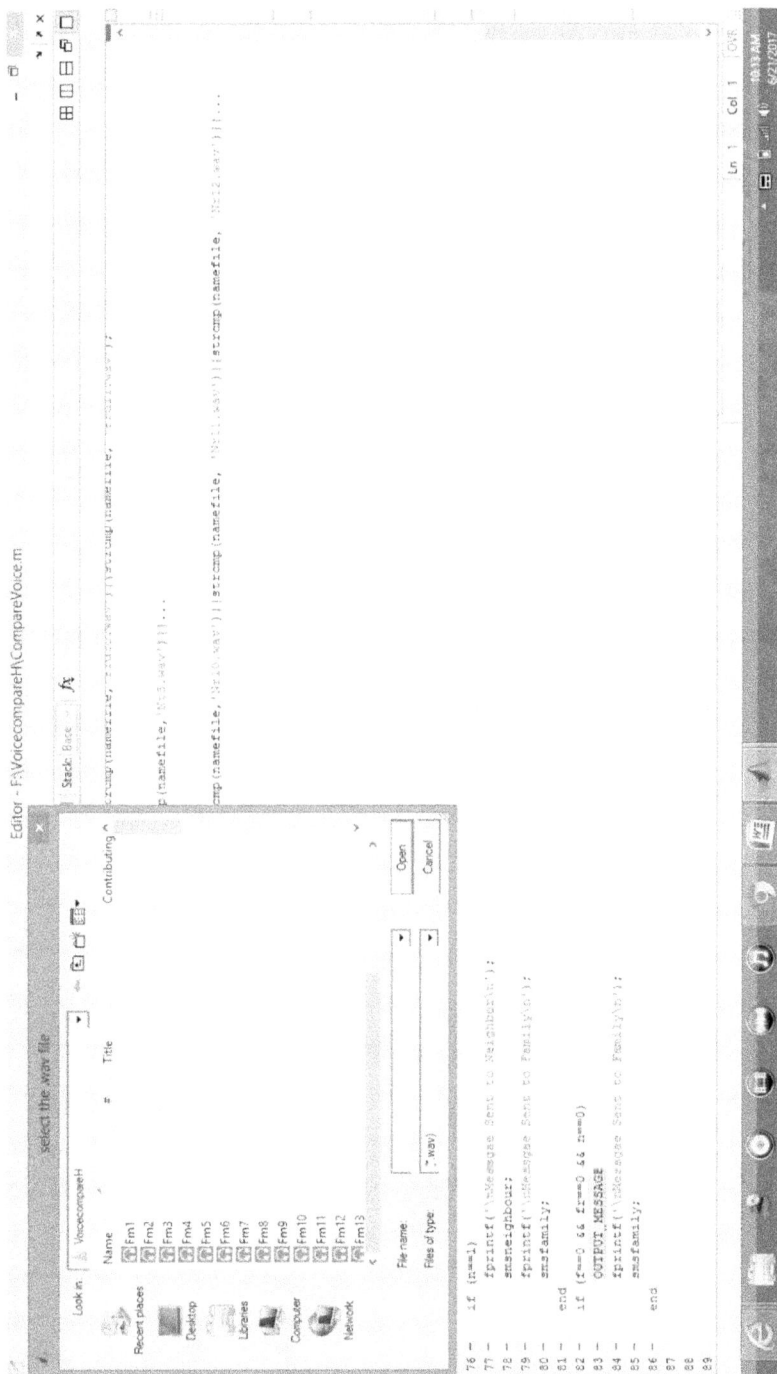

Figure 3.16 Testing of voice being done on the basis of samples stored in the database.

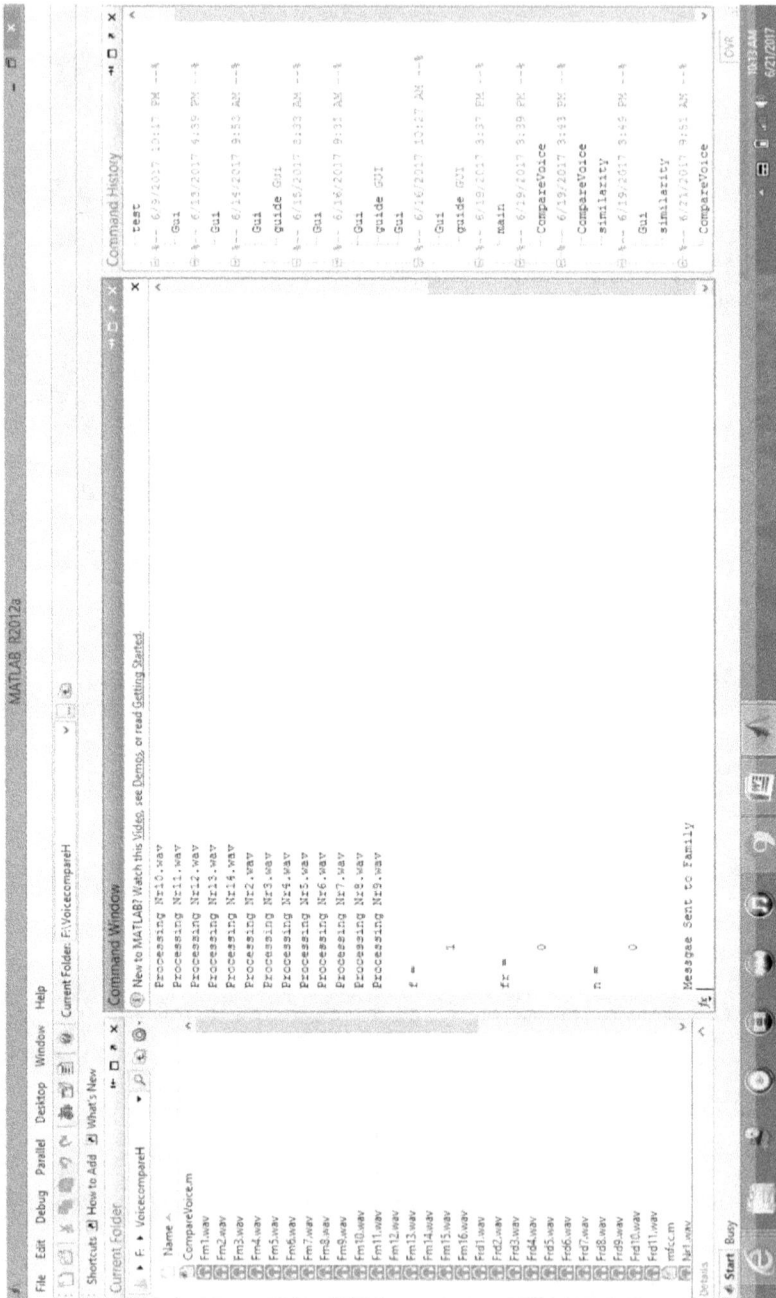

Figure 3.17 Result in voice testing based upon content matching.

The voice of the person at the door is captured and stored in an audio file and the features extracted from this audio file are stored in a matrix form in the database. Features of audio files are calculated by using the cepstral coefficients formula, i.e., ceps1-ceps2. The difference between cepstral coefficients of recently obtained audio file and each audio file already stored in the database is obtained. After subtraction, the mfcc value is calculated that decides the category to which the person at door belongs and stored in a variable t.

Figure 3.18 shows the matrix created to calculate the recently obtained audio file cepstral coefficients named as ceps1, and Figure 3.19 shows the matrix form of the each prestored audio file cepstral coefficients named as ceps 2. These values are stored in a matrix in the form of rows and columns.

After subtraction, variable t will provide the results in another matrix as shown in Figure 3.20, and finally the category to which the person belongs is displayed on the screen as previously shown in Figure 3.17.

Text Independent Voice Recognition (Similarity Index): This technique is taken into consideration when any one of two above mentioned and used techniques that are face recognition and voice recognition technique fails to recognize the person correctly or accurately. It is a text-independent voice recognition technique used to recognize voice based upon voice features such as pitch, speaking rate, voice modulation and frequency, irrespective of the text or content spoken by the person at the door. The voice features of person at door are compared with voice features of persons belonging to all four categories. If the difference between the features of voice of the person at the door and the features of voice of any of the stored voices is less than or equal to 0.014, voice is matched else that person is identified as belonging to unknown category as shown in Figure 3.20. This value considered for Similarity index, i.e., 0.014 is considered as threshold value for recognition. If similarity index is greater than or equal to 75%, then only the person at door will be declared as known person and the category to which he/she belongs will be displayed on the screen and depending upon the category he/she belongs, a text message will be sent to the concerned persons belonging to family, friend and neighbor categories and if the value of similarity index is less than 75%, no message would appear at the bottom of the window in MATLAB since the person at door belongs to unknown category.

Figures 3.21 and 3.22 show the output based upon values of similarity index. Figure 3.22 depicts a person as belonging to the category 'Unknown,' and Figure 3.23 depicts a person as belonging to category 'Family,' respectively.

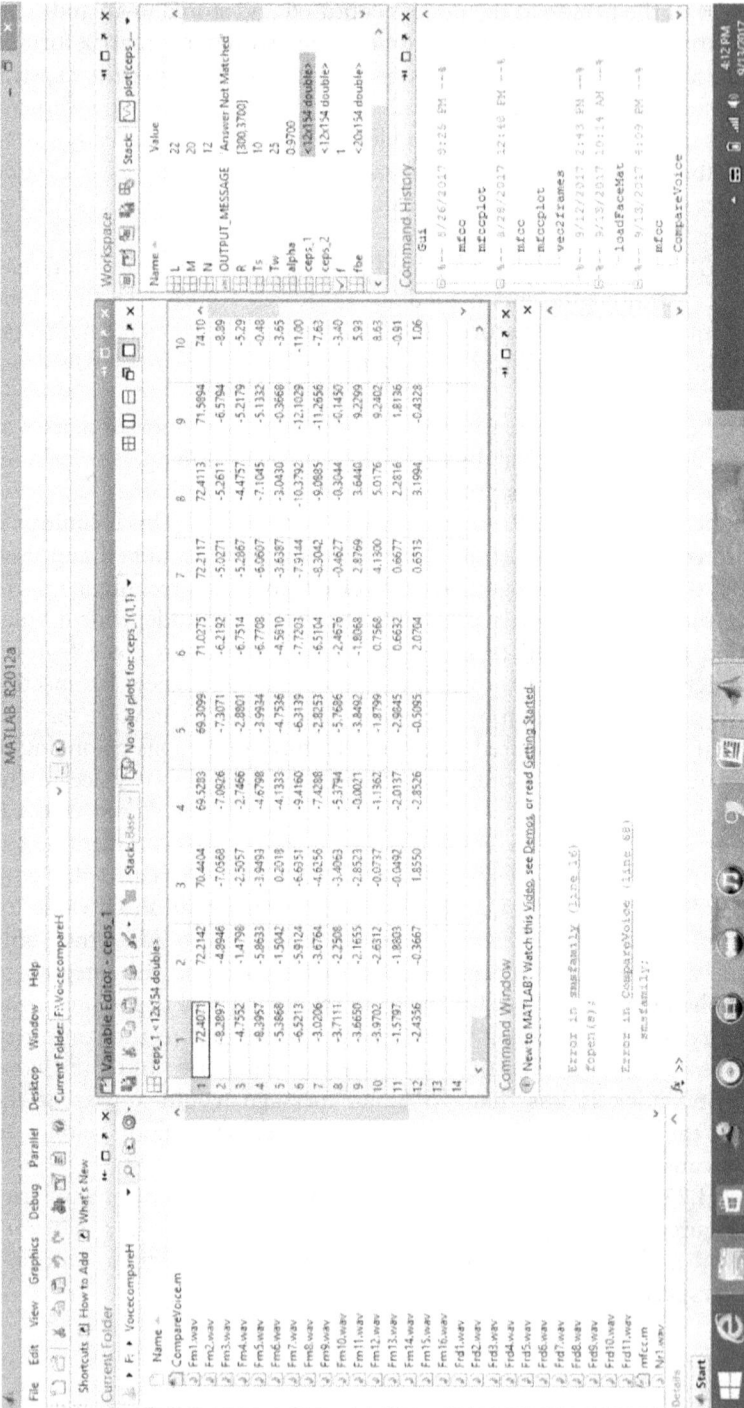

Figure 3.18 Matrix form of cepstal coefficients.

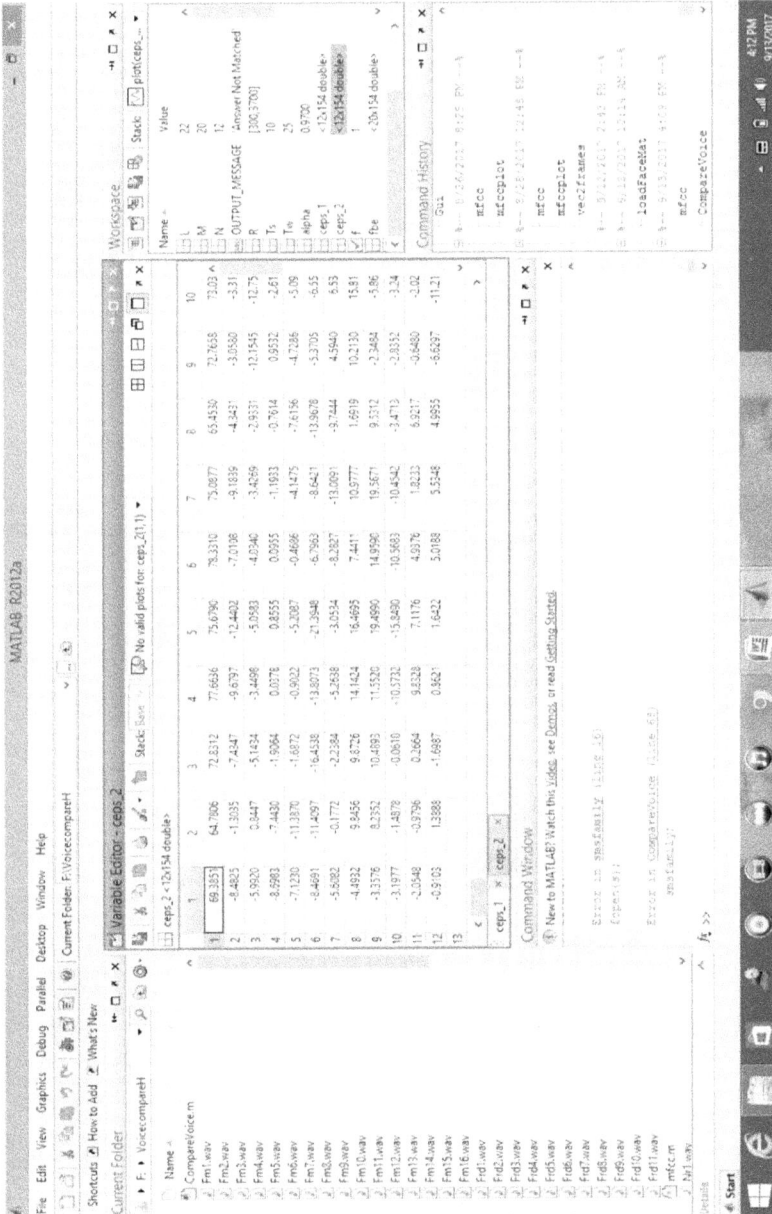

Figure 3.19 Matrix form of second cepstral coefficients.

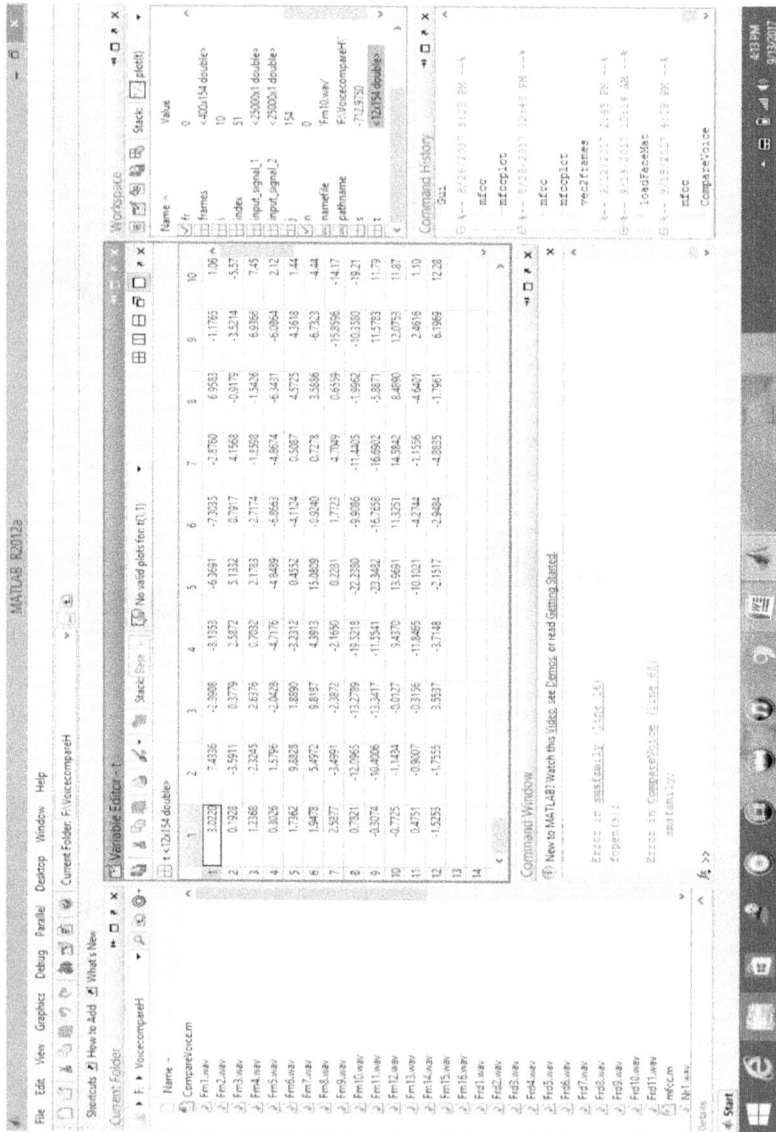

Figure 3.20 Result in a matrix form.

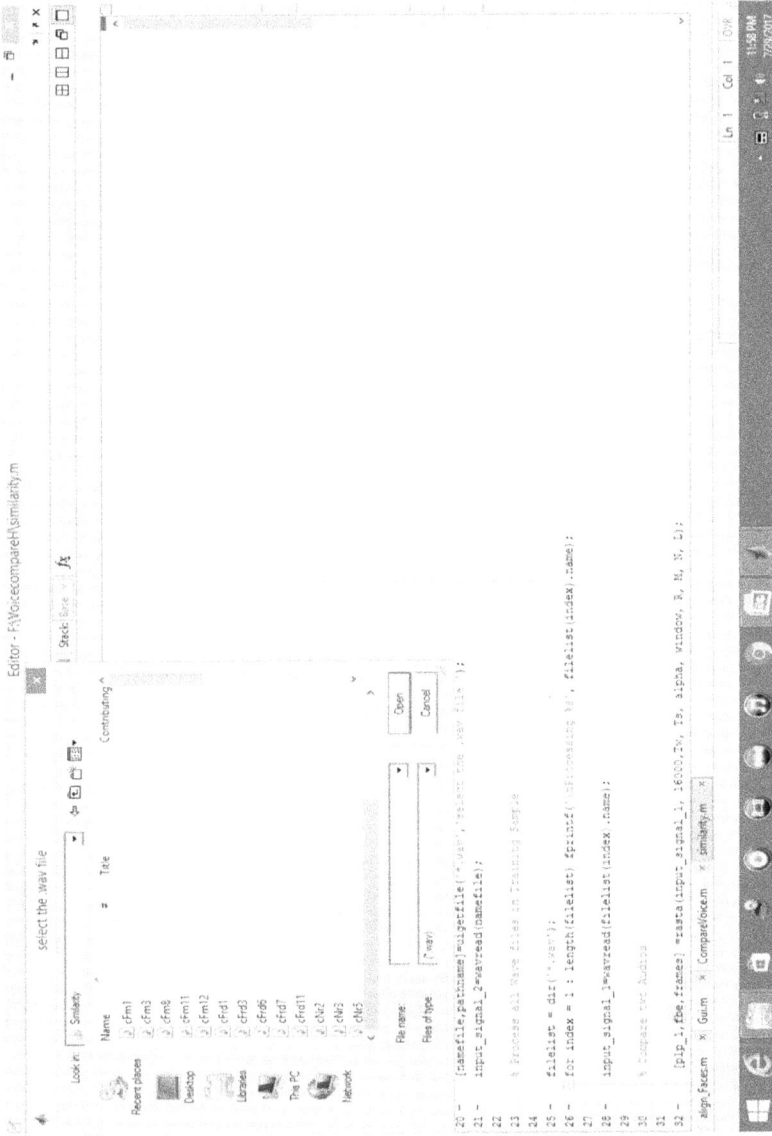

Figure 3.21 Similarity index testing based on voice frequency patterns.

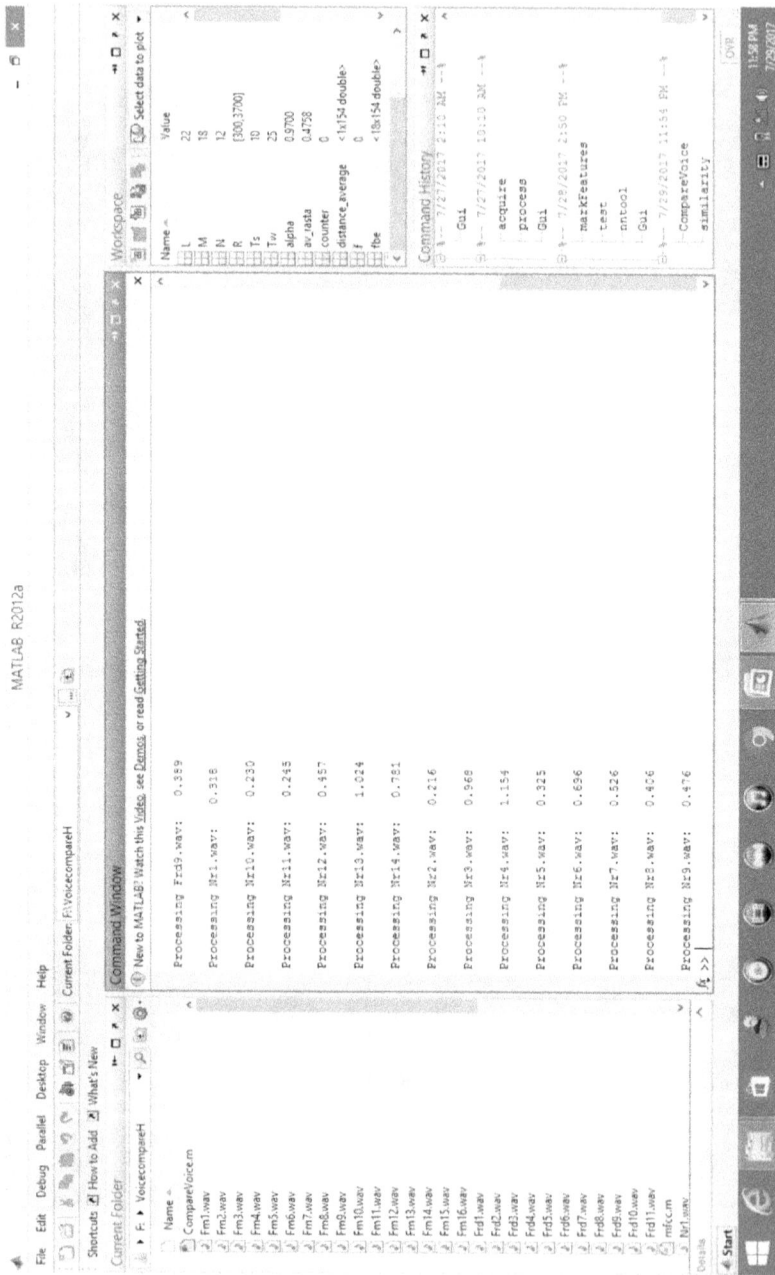

Figure 3.22 Result of similarity index for person belonging to "unknown" category.

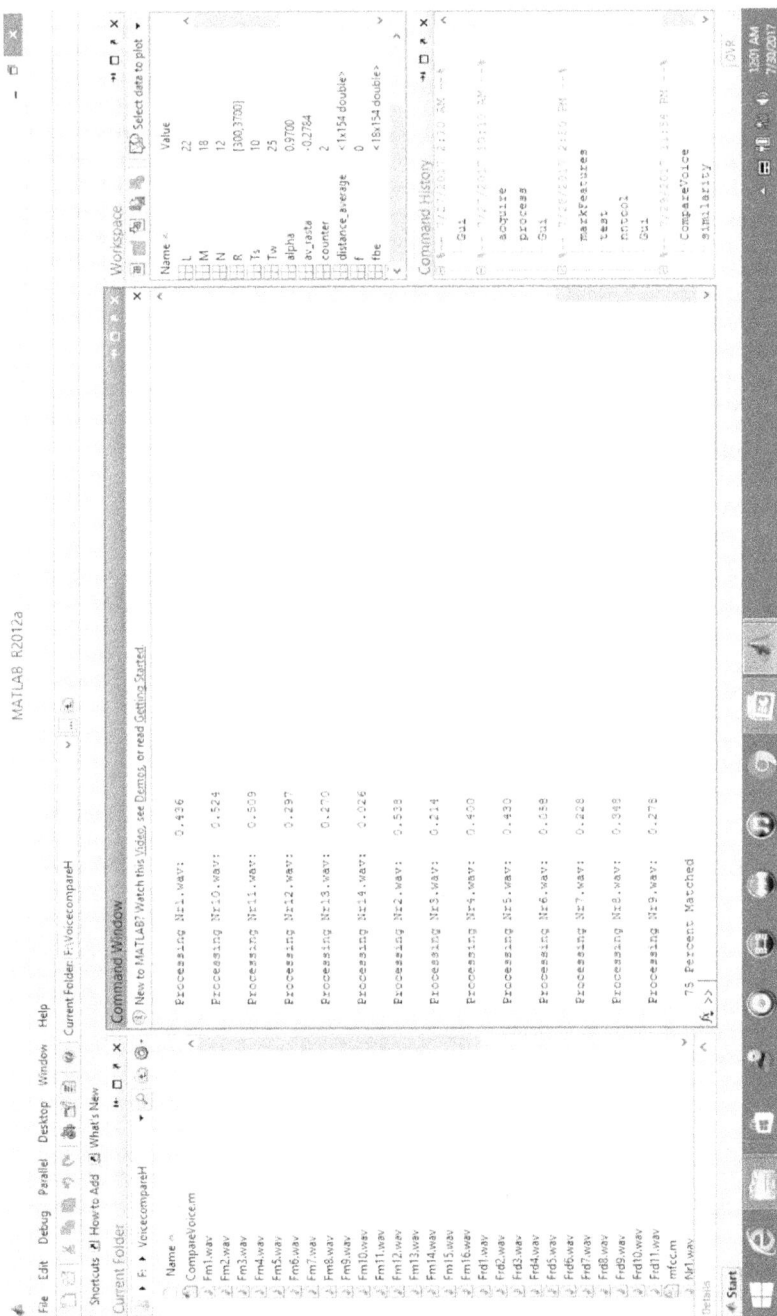

Figure 3.23 Result of similarity index for person belonging to "family" category.

3.7 RESULTS AND DISCUSSION

In this section, the proposed hybrid MFRASTA technique has been compared with the individual MFCC method and the comparison of MFRASTA technique with the individual MFCC method in terms of recognition rate by taking 30–50 audio files into consideration has been done as depicts in Table 3.2. Table 3.3 represents the percentage of accuracy obtained in case of the similarity index by taking a set of 15 known and 15 unknown persons. It is summarized that the proposed hybrid technique surpasses the previously used individual algorithms or techniques.

Figure 3.24 shows the performance graph of individual MFCC in terms of recognition rate.

Figure 3.25 shows the performance graph for the proposed hybrid MFRASTA technique along with individual MFCC by taking a parameter of recognition rate.

Figure 3.26 shows the performance graph of the similarity index for individual MFCC in terms of recognition rate.

Figure 3.27 shows the correlation of the proposed hybrid MFRASTA technology with individual MFCC for similarity index in terms of percentage of accuracy which shows 73% recognition rate for known persons and the 93% recognition rate for unknown persons.

Finally, messages are sent to the concerned persons belonging to concerned categories through a dongle preregistered number as shown in Figure 3.28. A number of persons belonging to different categories can receive the message at the same time.

Table 3.2 Comparison of proposed hybrid MFRASTA with individual MFCC in terms of recognition rate

No. of samples	MFCC	MFRASTA
I(30)	90	99
II(40)	90	99
III(50)	92	99

Table 3.3 Accuracy table for similarity index of known and unknown persons

Samples of persons	No. of samples	Correct answer	Incorrect answer	Accuracy
Known	15	11	04	73
Unknown	15	14	01	93

MFCC

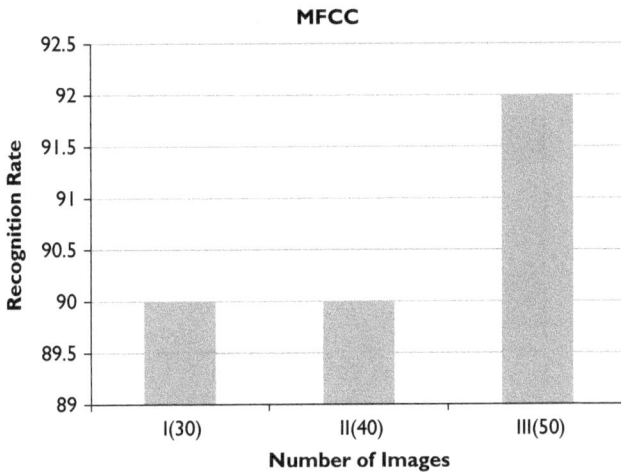

Figure 3.24 Voice testing based upon content matching.

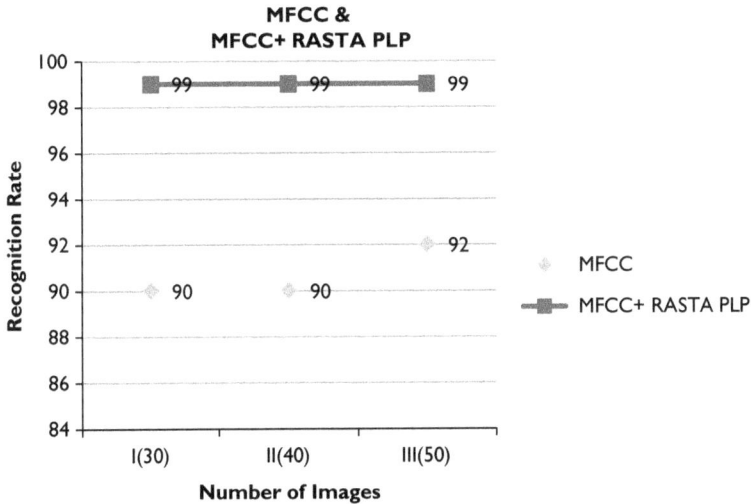

**MFCC &
MFCC+ RASTA PLP**

Figure 3.25 Performance graph for proposed Hhybrid MFRASTA technique and individual MFCC in terms of recognition rate.

3.8 SUMMARY

Complete security is needed at homes, offices, banks, etc., for providing the safety to the users. So, in order to provide security for an individual who is living alone at home, voice recognition technique with Similarity Index along with the face recognition technique explained in the previous chapter has been used. There are three main tasks involved in a voice recognition

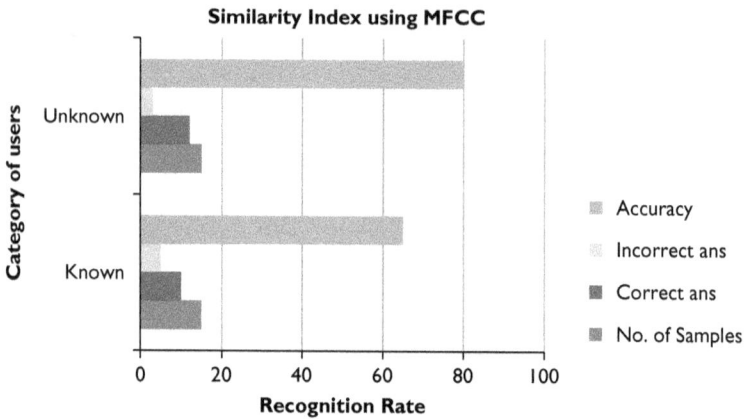

Figure 3.26 Recognition rate graph for similarity index using individual MFCC.

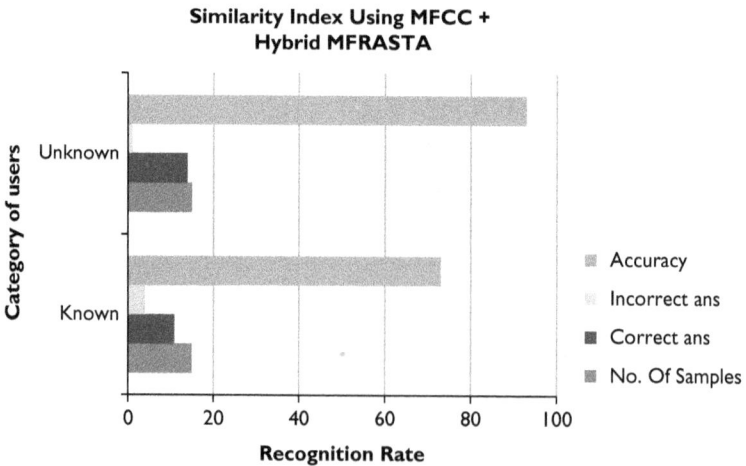

Figure 3.27 Comparison of recognition rate between individual MFCC and hybrid MFRASTA technique in terms of recognition rate.

system named as preprocessing, extraction of features and recognition of features. The purpose of taking out the features is to extract the features from voice signal and then transform them into feature vectors. Each voice recognition system comprises with two phases known as: enrollment phase and matching phase. First, in the enrollment phase, a voice model is maintained by using the feature modules of the person belonging to one of the three categories, i.e., family, friend or neighbor. In the matching phase, the feature vectors separated from the unknown person's sample of speech are compared with all samples stored in a database. This gives a value of similarity score. The last decision module uses this value of similarity score

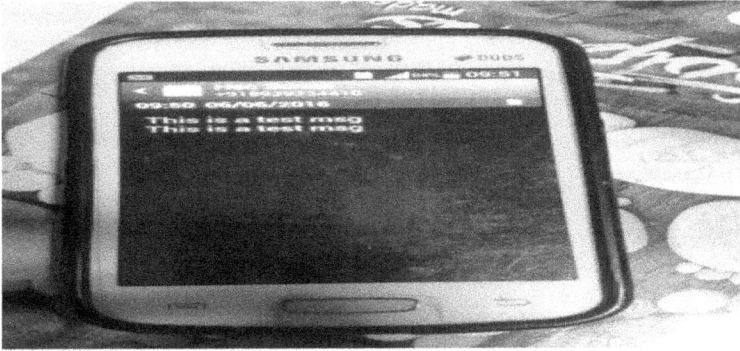

Figure 3.28 Screen shot of text message being displayed on cell phone.

to take the final judgment. The proposed system uses a hybrid combination of MFCC (Mel Frequency Cepstral Coefficient) with RASTA-PLP (Relative Spectral-Perceptual Linear Prediction) for the extraction of features and for classification of data. The motive of this proposed research work is to implement a voice recognition system, which will work for both text dependant as well as text independent recognition technology. The hybrid MFRASTA Technique has been implemented in this proposed work to obtain improved results in terms of recognition rate. This system is cost effective, highly reliable and provides accurate response. In future, security measures may be taken into consideration while taking and storing the audio files of the person and embedded sensors can be integrated in the proposed model to providing safe and comfortable environment. A system can be developed for more vocabulary size and accuracy can be enhanced by using speech enhancement technologies. The experiments carried out in this proposed work are based on speech samples taken in a noise-free environment. But in real-time conditions, a person may speak in noisy environment also. So, the effectiveness of the proposed work can further be extended by taking this factor into consideration.

Chapter 4

Sensor-based health monitoring system uses fingertip application

4.1 INTRODUCTION TO SENSORS-BASED SERVICES

Vital signs are the important functions that can be measured for a person. Each vital sign is measured by using a number of equipments. In today's society, there are a number of diseases rising very high. Heart failure and heart stroke origin a big burden on society due to the high cost of care, lower life quality and premature death [68]. So, there is a need to monitor some essential parameters such as BP, temperature and heartbeat by cost-effective devices for further treatment. In today's world, a number of ubiquitous devices are used named as laptop, PDA, tablets, smart phones, etc. A smart phone is one having some advance features except making calls and sending text messages. Users can simply use a smart phone having all the computer features on it by simply touching or swiping the screen. The use of communication technology and information when combined with engineering and medical concepts enables researchers in healthcare to increase patient monitoring at home, park, public places, hospitals, etc. In hospitals, they are provisions for monitoring the patients, but every time it is not possible to go to hospital and in that case, it's quite hard to handle that situation. Children, Adults and old age persons can take the benefits of healthcare technology by using customized monitoring. A number of methods are there to measure health parameters, for example, PTT (pulse transit time) for BP estimation, but these methods require high complexity devices that are not convenient for users to wear [69]. Heart rate and variability in that rate are two important factors which provide information related to diseases. Heart rate variability also indicates about strain and anxiety. Previous technologies such as heart rate sensor, finger pulse amplitude and electrocardiograms (ECG) are not applicable now [69]. Heart rate represents how many times a heart beats per minute (bpm). Previously, phonocardiograph's devices were used with equipments such as microphone, amplifier and filter to record sound of the heart.

There are a number of methods to measure heart rate, but most of the methods are uncomfortable. Figures 4.1, 4.2 and 4.3 show the traditional

DOI: 10.1201/9781003120933-4

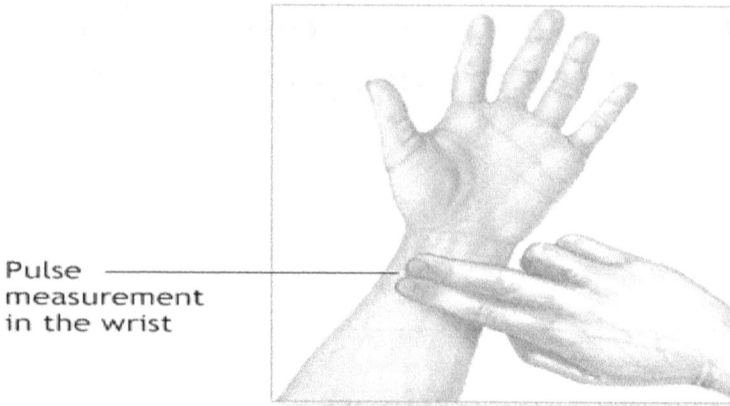

Figure 4.1 Manual way to measure the pulse.

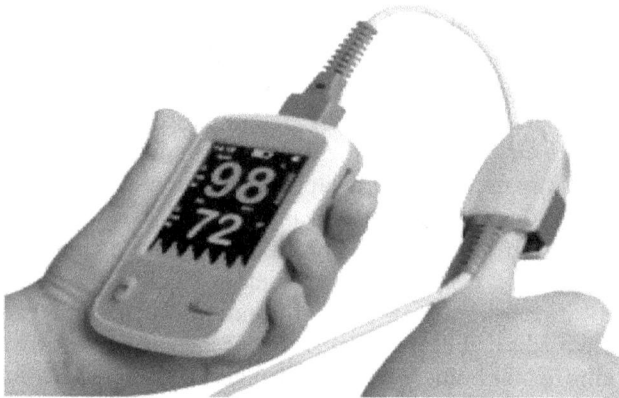

Figure 4.2 Hand-held pulse oximeter.

devices used for pulse measurement: a) manual way of pulse measurement, b) hand-held pulse oximeter and c) finger clip bluetooth oximeter heart beat monitor device.

Software applications extract vibration rate from sound of heart documented with microphone. The accelerometer is needed to check chest vibration produced by heart's movement. The accelerometer is having the ability to detect acceleration through MEMS (Micro Electro Mechanical System). Traditional methods include GPS (Global Positioning System) facility in order to track a patient's position [70], but it is highly energy consuming and indoor unavailability device. GPS is a satellite system that gives time and location related information to the receivers of GPS. Another method is wearable sensor-based devices such as a wristband, chest band and belt with several sensors on the device seems uncomfortable and highly irritating.

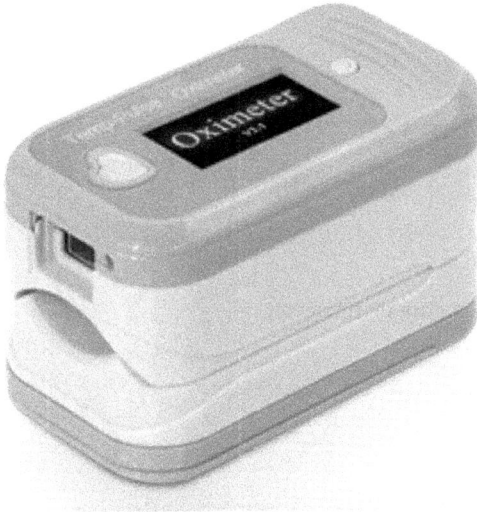

Figure 4.3 Finger clip bluetooth oximeter.

When the heart is beating, it is actually pumping blood all over the body and that makes the blood volume inside the finger artery to change too. This fluctuation of blood can be detected through an optical sensing mechanism placed around the fingertip. Heart rate measurement is also possible by using pressure sensors, light sensors and sound sensors. First is the pressure sensor that is used to detect the changes in pressure near the vibrations produced by the heart. Mainly researchers use this method to observe resting heart rate. This sensor is placed under the quilt and then changes are detected in the heart rate. It is a most popular way of heart rate measurement, which is mainly used when the user is in lying state by using sensor closest to the heart. But the drawback with this type is the sensitivity of the noise factor. The sound sensors are used to measure the sound changes near the heart in which air conductive microphone is used to measure the heartbeat parameters without direct contact with the skin and light sensors is used to find the changes in optical property of the blood. This method uses LED in infrared region and photodiode, which helps to detect changes in volume of blood.

Body's temperature is measured by the difference between the heat generated and lost. Number of times it is acceptable to measure body temperature in the mouth, ear and skin. Figure 4.4 represents mercury in a glass thermometer, Figure 4.5 shows an ear thermometer device that is very reliable and compact and Figure 4.6 shows wireless remote temperature measurement device. Number of factors is there that affects the accuracy of measurements such as cold or warm hands or clothes hot or cold fluids. Three figures given below show the different devices used for measuring body temperature.

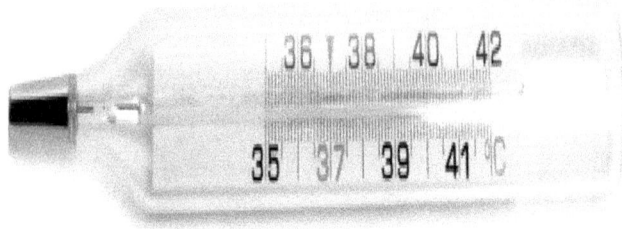

Figure 4.4 Mercury in glass thermometer.

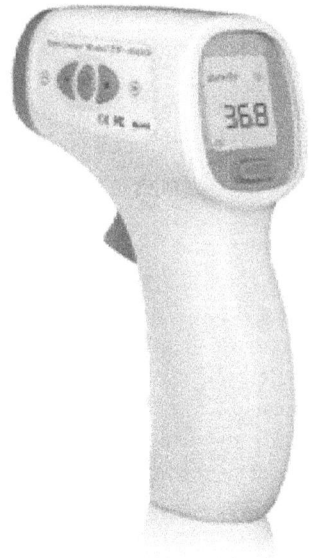

Figure 4.5 Ear thermometer measurement device.

Blood pressure is one of the important indicators of body of human being. In early 19th century, first time BP was measured by using a simple device [71]. After then, there has been a number of advancements in BP measuring devices. These figures depict the traditional devices used for measuring blood pressure: Figure 4.7 shows manual sphygmomanometer, which uses a stethoscope to measure the heart rate, in Figure 4.8, Boso-medicus device was there which was a wireless Bluetooth device and in Figure 4.9, automatic wrist heart beat monitor device is shown that was having large memory for the recordings.

Measurement of BP is measured in mmHg units. There are two types of blood pressures: systolic and diastolic. Systolic means to measure the maximum pressure pumped out by the heart and diastolic means minimum pressure where the heart relaxes. Too high or low BP may be dangerous to health.

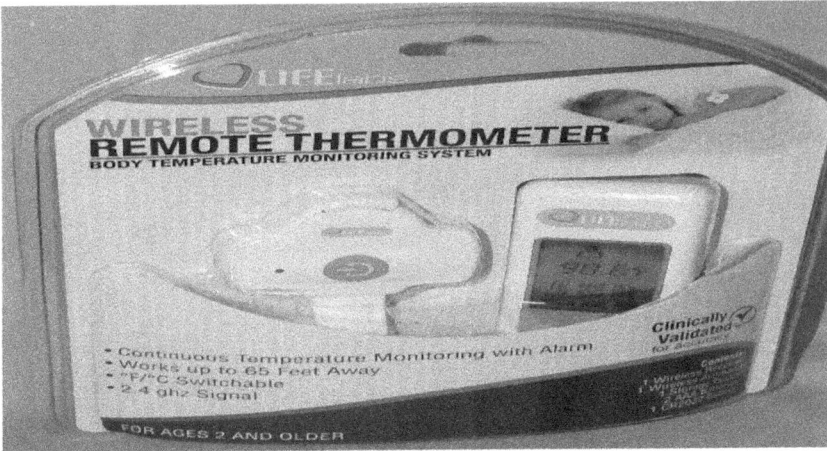

Figure 4.6 Wireless remote body temperature device.

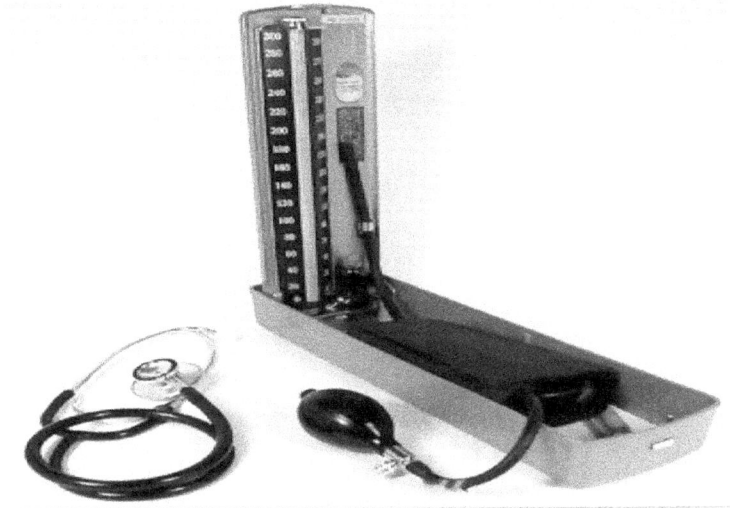

Figure 4.7 Manual way to measure the blood pressure.

Table 4.1 given above shows the different categories of blood pressure. In the present scenario, two new representations of care are emerged named as: self-care with home-based services. Advancement in sensor technology with network technology has made this task available [128]. The sensor unit is made up of an infrared light-emitting diode that transfers an infrared light into fingertip and a photodiode which senses the portion of light that is reflected back. Heart rate of healthy adult at rest position is around 72 bpm and babies is around 120 bpm while older children have around 90 bpm. As the elderly population is growing with increasing life expectancy, the

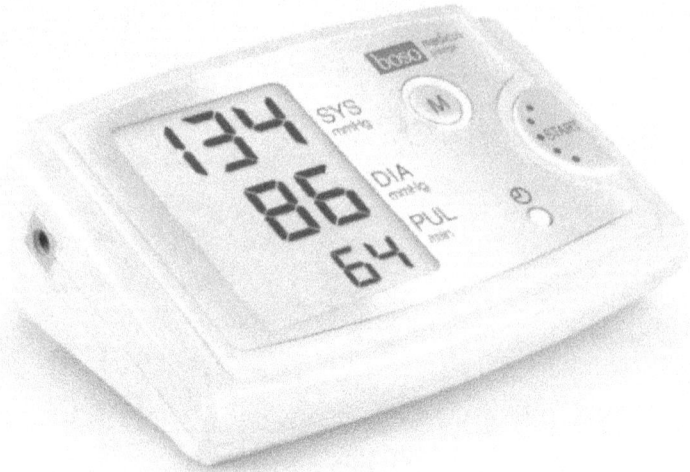

Figure 4.8 Boso-medicus BP monitor device.

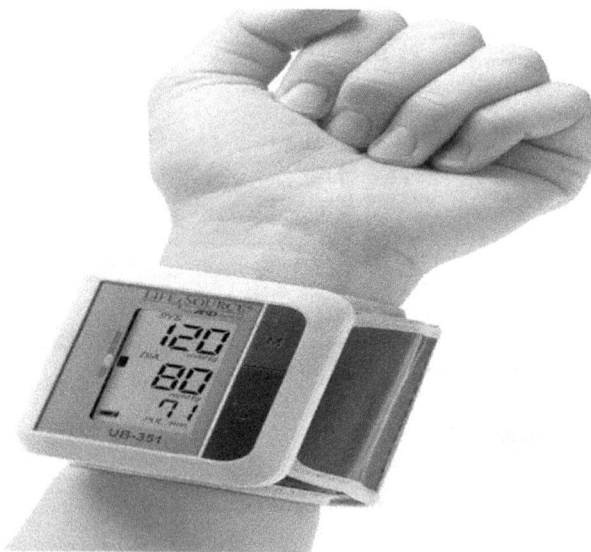

Figure 4.9 Automatic wrist BP monitor device.

number of challenges has been brought too many aspects of human life, especially in health monitoring, security and comfort level. The care of the elderly could be enhanced through health-monitoring system, sensor technologies and communication systems. Elderly/disabled persons or women who are living alone can take the benefits of this monitoring system and emergency system in order to maintain healthy and safety living [71].

Table 4.1 Categories of blood pressure

Category	Systolic	Diastolic
Hypotension	<90	<60
Normal	90–120	60–80
Prehypertension	121–139	81–89
Stage 1 hypertension	140–159	90–99
Stage 2 hypertension	>=160	=100

In previous time, persons are going to the hospital for heart monitoring where a cardiologist or officer will examine the patient for any heart diseases. ECG machine generally found in big hospitals and requires a specialist to handle the machine. Development of a device for heart beat monitoring with low cost using infrared sensor will be the foundation.

Architecture must be developed in a way to collect the data, analyze that data and then make it accessible by keeping the physician up to date using important signs and values which are checked by the interaction with physician and finally emergency call is initiated. This is very important that the technology used must be easy and transparent to the user. Thus, a reliable emergency system must be there to monitor the condition of the elder and in case of emergency, immediately inform the family members about the incident and call the ambulance as well as doctor [72]. So, the aim is to overcome the problems of any one's life as long as possible, so that patients can live an independent life in a better way. The main objective is to design and develop a low-cost portable heart beat monitor device using infrared sensor. The prototype that is generated is a device of low cost which can easily detect and shows the heart rate by counting the pulse generated using infrared sensors. In today's world, doctors, family members, friends and neighbors can easily receive health-related messages of their patients on their cell phone by using a MATLAB® tool and a 3G dongle. Monitoring of patients is playing an important role in taking the decision. This system is expanding now days by using its advanced features and capability to convert data into useful information. It is noticed that an ideal monitoring system should be able to gather quality data using sensor based devices, present that data in a meaningful way, perform decision support actions by using expert knowledge and finally take appropriate decision on the basis of collected data.

4.2 PREVIOUS TECHNIQUES

4.2.1 Phonocardiograph application

Phonocardiogram records the sound made by heart with a machine named as Phonocardiograph as shown in Figure 4.10. This is a medical device that

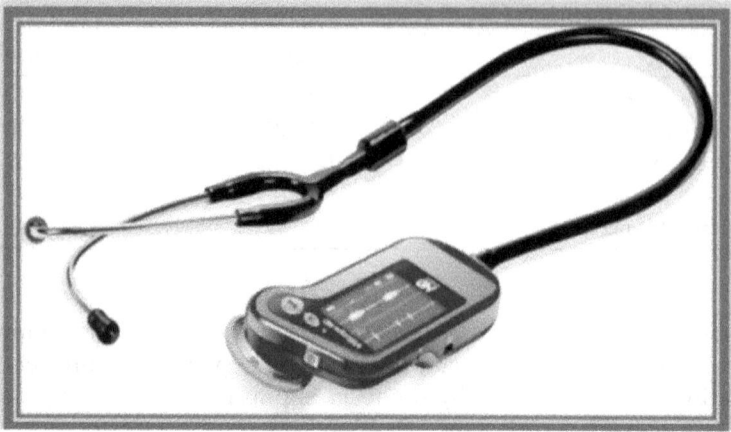

Figure 4.10 Phonocardiograph device.

uses number of techniques such as filters, microphones and amplifiers in order to collect heart sound. Integrated microphone is used to record the heart sound and then software applications are used to extract the pulse rate from that heart sound. But with this device, measurements are not accurate and it is very sensitive to noise. Opening and closing of valves in the heart produce sound during contraction and dilation, which is normally audible through a stethoscope. These sounds are rhythmic to heartbeat and can be sensed using microphones. Normal heart sound is used to determine the heart rate.

4.2.2 Mercury sphygmomanometer

It is a more iconic device found on every doctor's table, developed by Stephen Hales. This word is a combination of two words: sphygmos which means pulse and manometer which means pressure meter. This device is used to calculate blood pressure as it is possessed of an inflatable cuff to collapse and then releases the artery under cuff in a controlled way and a manometer is used to measure the pressure. This device requires a cuff, bladder, tubing and rubber bulb as shown in Figure 4.11. All are required in a maintained condition and serviced at proper time according to the instruction of the manufacturer. It is a simple gravity based unit. When this device is well maintained, it provides absolute measurements of blood pressure.

This device is very delicate and special care should be taken while operating, storing or transporting the unit. The advantage of using this device is it is very durable, it does not require readjustments but problems with this device is it is a bulky device that should be handled carefully to prevent damage. Errors may develop when it is not kept vertical.

Figure 4.11 Mercury sphygmomanometer.

4.2.3 Automatic digital sphygmomanometer

This device is known as oscillometric device, as shown in Figure 4.12. It uses an electronic pressure sensor for measuring the blood pressure, and readings are given out digitally on the display. This device has inflatable cuff like mercury sphygmomanometer and cuff is attached to the electronic unit. Difference with other types is the technique used. Its report is based on the sounds produced by blood, flowing inside the arteries. It evaluates and measures the oscillations of the arteries using pressure sensors. As the cuff is inflated and deflated, oscillation occurs. These oscillations are processed using an algorithm to produce systolic and diastolic values that are digitally displayed on display screen.

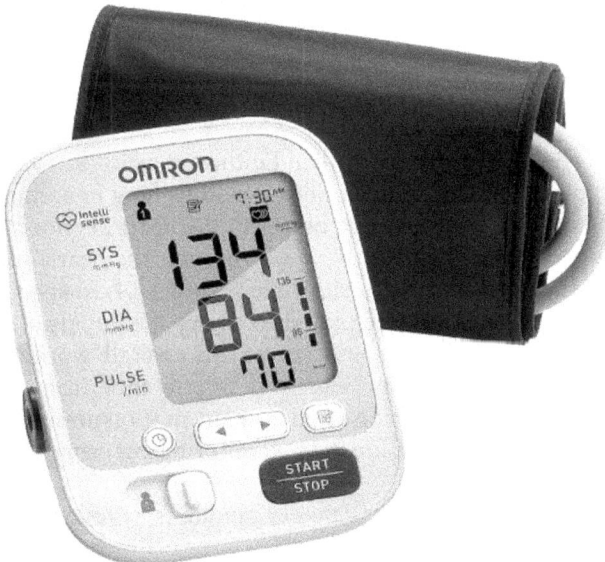

Figure 4.12 Automatic digital sphygmomanometer.

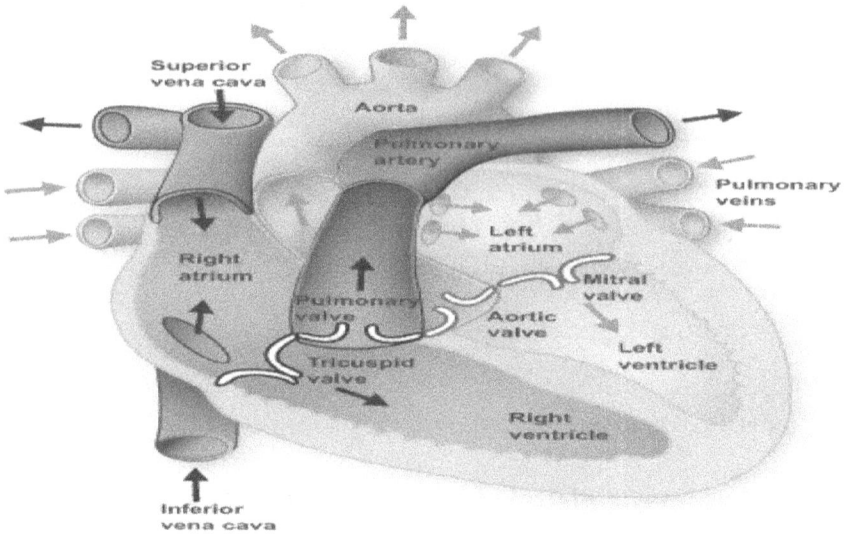

Figure 4.13 Anatomy of human heart.

They are battery operated. The advantage of using this device is that it is compact and portable. Operating the device is easy and chances of error are minimized, but drawback is it is delicate and proper care should be taken while handling the device.

4.2.4 Electrocardiographs

As the heart is four-chambered muscle whose function is to pump blood throughout the body. The heart is really two 'half hearts' left and right which beat simultaneously as shown in Figure 4.13. Each of these two sides has two parts or chambers: upper chamber known as an atrium and lower chamber known as ventricles.

This term was introduced by Willem Einthoven in 1893. The contraction and relaxation of cardiac muscles cause blood to flow in and out of heart. During each cardiac cycle, a set of tissues in the heart develops impulses in electrical form and then spread them all through the heart, which causes shrinkage of the heart muscles as shown in Figure 4.14. At specific points in the human body, electrodes are placed to detect the impulses. Then, ECG acquires this varying impulse and shows the overall working of heart. Typical ECG is corrupted by: electrical interference from surrounding equipments, measurement of noise and instrumentation noise. The ECG is a traditional method for heart function monitoring by using electrode contracts attached to the body.

Formula for calculating heart rate per minute is as follows:

$$\textit{heartbeatrate}/\min\ = 1/[\textit{pulseperiod}/(3 * 512 * 60)] + 92160/\textit{pulseperiod}$$

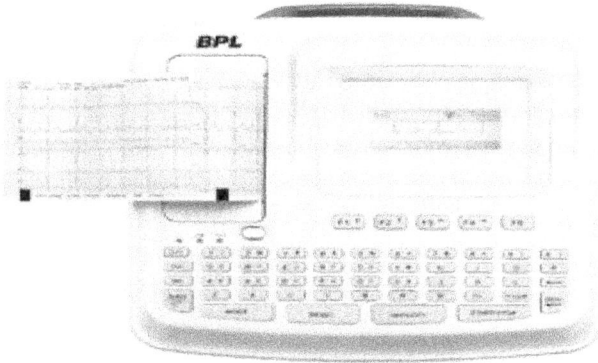

Figure 4.14 Electrocardiographs (ECG).

4.2.5 Accelerometer-based application

This application can be used to calculate vibrations of the chest caused by the movement of the heart. This method is easily affected by noise and gives inaccurate results most of the time. It detects heart rate without touching the body, based on Ballistocardiography (BCG) principle. Its benefit is a contactless measurement that enables continuous monitoring without disturbing the patient as shown in Figure 4.15. It is very easy to use.

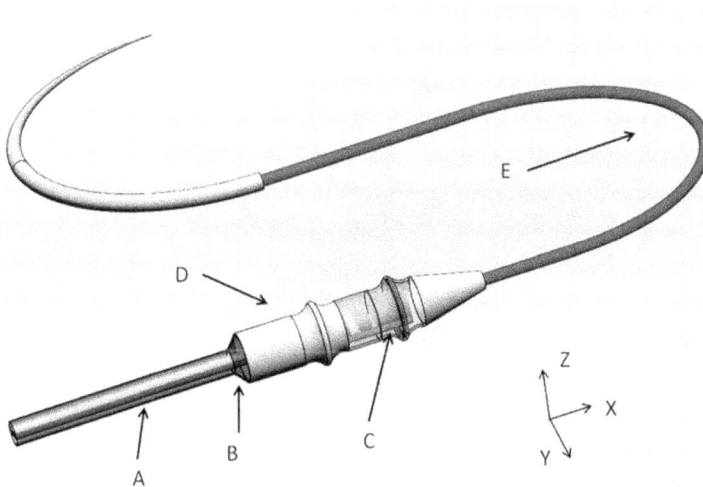

Figure 4.15 Accelerometer-based application.

4.3 PROPOSED TECHNIQUE: FINGERTIP APPLICATION

The proposed technique is based on a smart phone for health monitoring by using the concept of sensors. Smart phones are carried by everyone in every place and at all the time, this all time available property of having the smart phone makes it perfect to monitor health parameters and in an emergency, send text message to the family members. As the purpose of the proposed research work is to implement a low-cost portable device for health monitoring using infrared sensors and in order to achieve this objective two modules are used: Acquire and Process. In an acquisition, sensor in smart phone is used to acquire the waveform using fingertip for 5–6 seconds. This waveform consists of a number of frames and by using the concept of sampling and Fast Fourier Transform algorithm, extracts the value of sampling rate. In the process, peak detection and smoothing are used to get clear results of waveform.

The first step is pressing the smart phone camera lens for 5–6 seconds, so that a reading is obtained as shown in Figure 4.16. When fingertip is used to cover the camera, the screen of a cell phone is not black, it is actually red and the flash is turned ON as proper amount of light can reach the finger for accurate measurement.

In this work, videos are recorded with an iPhone 6s and copied into the laptop in.mov format as this format is compatible with MATLAB tool. Then, this video is processed by using a MATLAB tool.

Steps of proposed work:

The complete steps for proposed application are mentioned in a block diagram as shown in Figure 4.17.

Video signal acquisition: Bandwidth is the first thing that is required when sampling the signal is done. Heartbeat of a human being in normal

Figure 4.16 Fingertip placed on the camera lens.

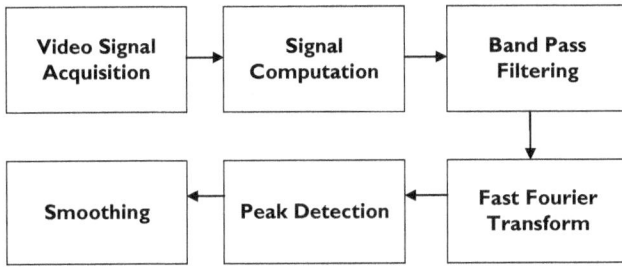

Figure 4.17 Steps of fingertip application.

condition is between 60 and 200 bpm (beats per minute), but it varies according to age, fitness and physical activities. In the proposed work, it is assumed that normal heartbeat can be found between 40 and 230 bpm. With iPhone 6s, videos are recorded at 87 frames as shown in Figure 4.18.

Brightness signal computation: Signal that is to be processed is a brightness of skin over time, but the brightness variation is not included in all the pixels of the image which is required, so choosing to collect all the pixels in a simple average brightness output value per frame. In proposing implementation, a part is skipped, which contains brightness of image computation by taking three planes named as: red, green and blue planes as the average of all the pixels is taken by using the single red plane. Red plane consumes all the energy so it provides similar results and it is cheaper too.

Band-pass filtering: When the signal is acquired, after that a filter is used named as: band-pass filter that works by disable the frequencies which are exterior to the interest band. The benefit of using this filter is that noise factor is reduced in later steps of processing and helps in making the results

Figure 4.18 Acquisition of signal in red plane.

smoother. In the proposed work, butter worth filter is used as it is an IIR type filter and with a given bandwidth, the order required is much less than as compared to FIR filter with cutoff frequencies set to 40–230 bpm.

Fast fourier transform: There are two types of transforms that are mainly used: Discrete fourier transform (DFT) is required to convert the signals from time domain to frequency domain, while FFT (Fast Fourier Transform) algorithm is required for recovery of processing time. When a comparison between DFT and FFT is done in terms of computational complexity of N points, DFT provides a value $O(N^2)$ and FFT gives the similar results having value of $O(N \log(N))$. It represents a high increase in speed in case of FFT.

Sliding window: Continuous heart rate estimation is done by using three methods: 1. FFT, 2. peak detection and 3. smoothing after each and every 0.5 seconds. The calculation is constantly implemented over a window implemented with last 6 seconds of signal samples. Basically, in this the window moves over the signal that is why it is known as sliding window. This window is moved at every 0.5 seconds because computing an estimate at this time does not rise the output's accuracy of time but time resolution is increased in case of reading.

Leakage reduction: DFT works excellently when time signals are infinite. A time limited signal having length N is correlative to multiplying its counterpart with a length N rectangular signal. This outcome in evolving the time signal spectrum of infinite length from which leakage is produced and to reduce this leakage, a function is used with boundaries zero. When input signal is multiplied by this function, the resulting boundary values will be zero. For that, Hann window is used as it offers good resolution and leakage rejection.

Peak detection: Once FFT is measured from the contents of current sliding window, the value of peaks in interest band is calculated by using a function named as 'find peaks' in MATLAB tool.

Smoothing: In order to smooth out the signals of heart beat readings more continuous, sliding window is coordinated with a number of phases having a series of tones in phase and finally smooth heart rate is taken by collecting frequency correlates to highest magnitude.

4.4 RESULTS

Proposed system behavior is evaluated by essential parameters, which are measured with an iPhone 6s camera and experiments are performed on individuals. This step is used to monitor the health parameters of the user. Here the video containing finger tip impression is browsed in order to calculate the sampling rate as shown in Figures 4.19 and 4.20.

Figure 4.21 provides a graph for heart beat calculation. It shows the value equals to 58 bpm. It means heart beat is in normal range. In case, the value goes up or down from the specified values, then only message will be sent to all family members.

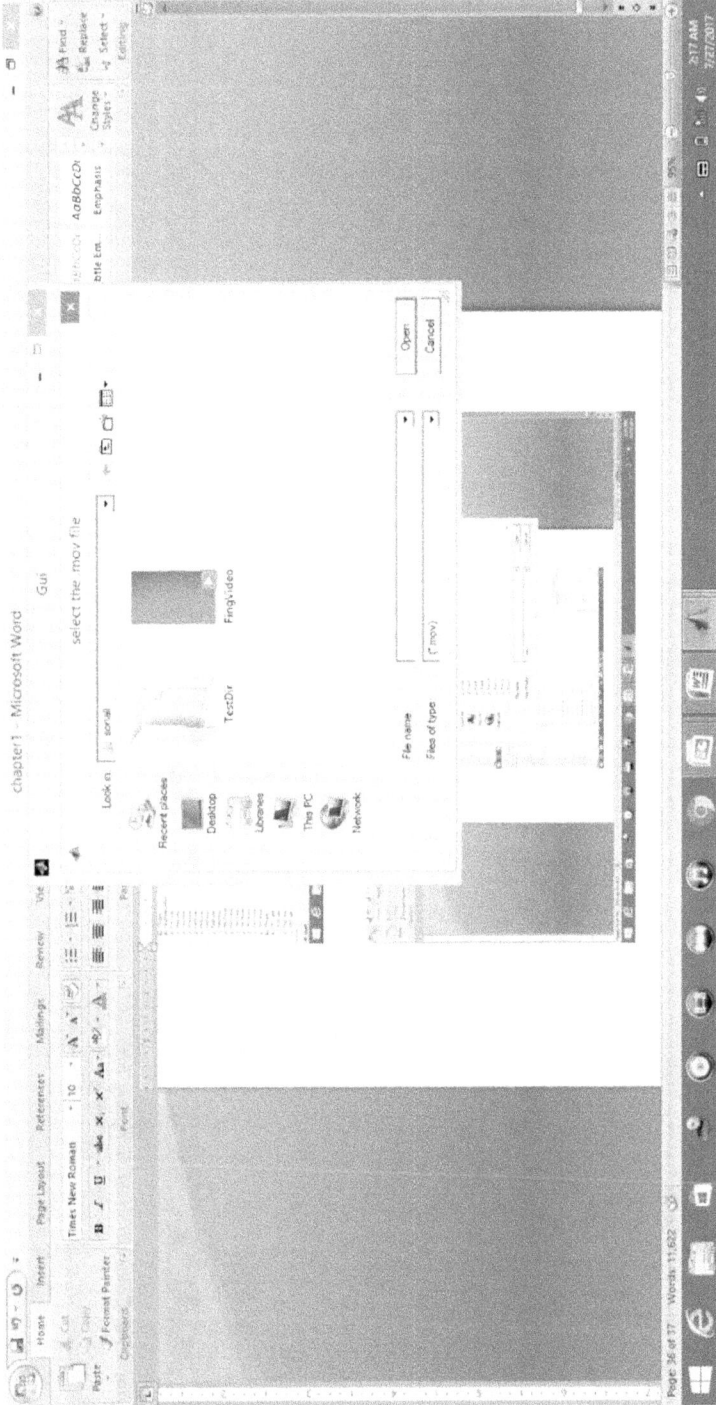

Figure 4.19 Health monitoring through fingertip impression.

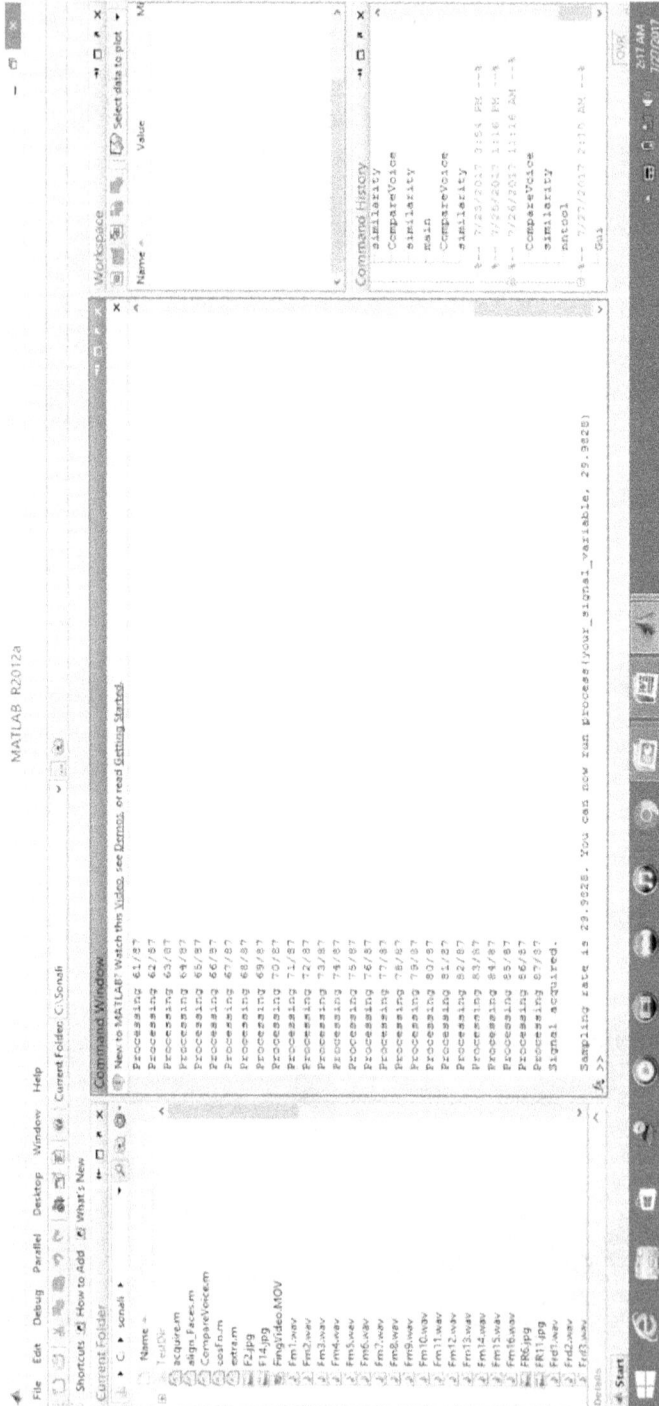

Figure 4.20 Calculation of sampling rate.

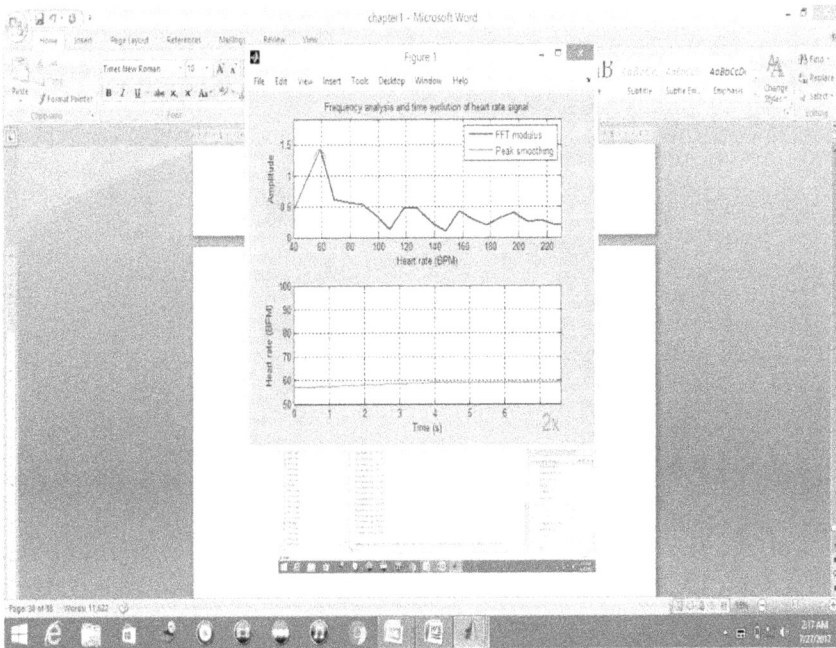

Figure 4.21 Graphs for heart beat monitoring.

Then, the message is sent to the family members from the number, which is registered with the dongle. If the value for heart beat is higher or lower than the threshold value, in that case, the emergency message will be sent [73]. Figure 4.22 shows the screenshot of message on phone received by using MATLAB tools and iPhone 6s. When proposed work is compared with previous work using home care devices, outcomes are better mainly in terms of false acceptance rate and accuracy. Advantages of using this proposed method are to get accurate results even with a low factor video camera that have decreased capture rate, it draws limited power and can be used regularly and it has low computational cost. It is opposing to noise and gives favorable results even with uncertain conditions.

4.5 SUMMARY

Everyone knows that the best way to deal with a critical situation is by taking a fast response. By using this current application in a way to get rapid alarms about critical situations that put the life of an infant at risk. In this proposed work, a number of approaches are studied to extract heart beat measurements but it is assumed that the suitable way to choose a monitoring device is without touching that device. A method of Fingertip application for smart phone users is successfully used in this work by

Figure 4.22 Screen shots of text message displayed on the phone.

comparing the results with previous methods. This method gives correct results even with low quality video cameras. This proposed method uses built in hardware, i.e., camera, integrated sensors and simple ways of communication with each other, e.g., sms, phone, email, etc. It is also examined that red plane is sufficient to obtain desired results irrespective of RGB planes and filter used is very lightweight, which is suited for devices with less resources. This method gives correct results for heart beat measurement. Persons can efficiently run this application when they rest at home, outside home e.g. in the park or during any activity. This application is suitable for all the users when they want to record their heartbeat on a daily basis. In future, some other health parameters may be considered like temperature, blood pressure, sweat rate, etc. This collected data can be used for simulating in a controlled environment and possibility of false alarm can be studied under nonoptimal conditions.

Chapter 5

Hybrid **PICA** and **MFRASTA** technology with sensor-based fingertip application for individual's security at home

5.1 PROCESS OF FACE AND VOICE RECOGNITION

When this application is run, first face recognition process starts. In face recognition, hybrid PICA algorithm is proposed and in voice recognition, the hybrid MFRASTA technique is proposed for obtaining the desired results. The combination of face + voice recognition technology overcomes the drawbacks of individual techniques. Each system will deliver the output in the form of the known or unknown user [74]. The objective of the proposed research work is to implement a highly secure system, which can be used in homes, offices, banks, etc.

Figure 5.1 depicts the architecture of the proposed model. In this model, an image will be acquired by a camera at first and a face recognition process will be applied on this image using an image acquisition toolbox available in MATLAB® tool. In the current scenario, images taken into consideration are of $512*512$ dimensions. A procedure may be used to convert the size of each received file from any other dimension to specify dimension with the help of any of the existing software. In the proposed model, four-step procedures for face recognition have been used.

a. A training set is taken and its preprocessing is done in which images are normalized using different methods such as PCA and DCT.
b. Independent features are extracted by using two face recognition algorithms PCA + ICA in combination [75], [76].
c. Feed forward neural network is preferred for classification which classifies a given input image.
d. The output of each neural network is combined to classify input image that defines the category of that input image to which the person at the door belongs.

This is a unimodal technique of biometric. If work is done on single authentication technique, it does not fulfill our requirements regarding to performance, noise problem, etc. So to overcome these issues, multimodal

DOI: 10.1201/9781003120933-5

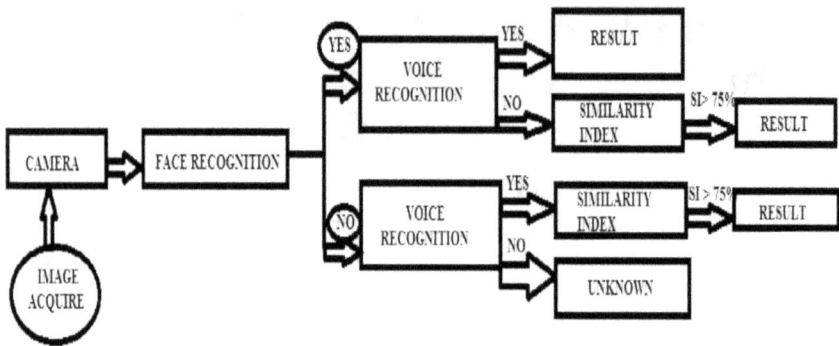

Figure 5.1 Proposed model architecture.

technique of biometric has been taken into consideration in the proposed model. A combination of two authentication techniques has been used: face recognition at first and then voice recognition in a single system. The main aim to combine these systems is to increase the performance and accuracy of the model.

After face recognition, steps are taken for voice recognition which is as followed:

- First step is to taken data in training set and apply preprocessing in which audio files are recorded by number of methods in.wav format.
- Feature extraction step extracts the features that provide specific information from the voice signals. Features have been extracted by using two voice recognition algorithms MFCC and RASTA-PLP [77], [78].
- Combine all the outputs and classifies input voice that it is known or unknown.
- Send the message to define the category of each input voice.

Four scenarios are considered here for the high security of the individual.

a. When face recognition and voice recognition both are successful
b. Face recognition test passes but voice recognition test fail.
c. The face recognition test fails but voice recognition test passes.
d. When face recognition and voice recognition both are unsuccessful.

In the first scenario, category of the person at the door is displayed as output on the screen out of any proposed categories: family, friend, neighbor or unknown. In the second and third scenario, similarity index technique comes into existence. This technique is taken into consideration when any one of two above mentioned and used techniques that is face recognition or voice recognition technique fails to recognize the person correctly or accurately. It is a text-independent voice recognition technique used to recognize voice based upon voice features such as pitch, speaking rate, voice modulation and frequency, irrespective of the text or content spoken by the person at the door. The voice features of the person at the door are compared with voice features of persons belonging to all four categories. If the difference between the features of voice signals of the person at the door with the voice features of any of the stored voices is less than or equal to 0.014, voice is matched else that person is identified as belonging to unknown category. This value considered for Similarity index, i.e., 0.014 is considered as the threshold value for voice recognition. If Similarity index is greater than or equal to 75%, then only the person at door will be declared as known person and the category to which he/she belongs will be displayed on the screen and depending upon the category he/she belongs, a text message will be sent to the concerned persons belonging to family, friend and neighbor categories and if the value of similarity index is less than 75%, no message would appear at the bottom of the window in MATLAB since the person at door belongs to unknown category. In fourth scenario, output displayed on the screen shows the person at door belonging to "unknown" category.

Figure 5.2 shows the GUI of overall proposed architecture that includes Face Recognition, Voice Recognition, Similarity Index and Vital Sign Monitoring. These are the same steps as explained earlier as there are total four cases in this technique as shown in Table 5.1. If face and voice recognition is done properly, then the message will be sent to all category members.

The face recognition process starts by selecting any image from the database as a number of images are stored there which belong to different categories (friend, family and neighbor). This process is implemented using MATLAB tool as shown in Figure 5.3. An image to be identified is matched with all the images that are stored in the database and when this identified image matches any of the stored images, the screen will stop for a few seconds and then it will display the category to which the image belongs as shown in Figure 5.4.

When face recognition process is done, the percentage will be displayed on the screen along with the category to which the person at door belongs as shown in Figure 5.5.

After face recognition, the voice recognition process will take place as shown in Figure 5.6.

As shown by Figure 5.7, output given by both face recognition technique and voice recognition technique is family category. So, that means image to be identified belongs to family category. Hence, a message will be sent to all family members and neighbors already registered in the database by using a MATLAB tool and 3G dongle.

The snapshots for similarity index are shown in Figures 5.8 and 5.9.

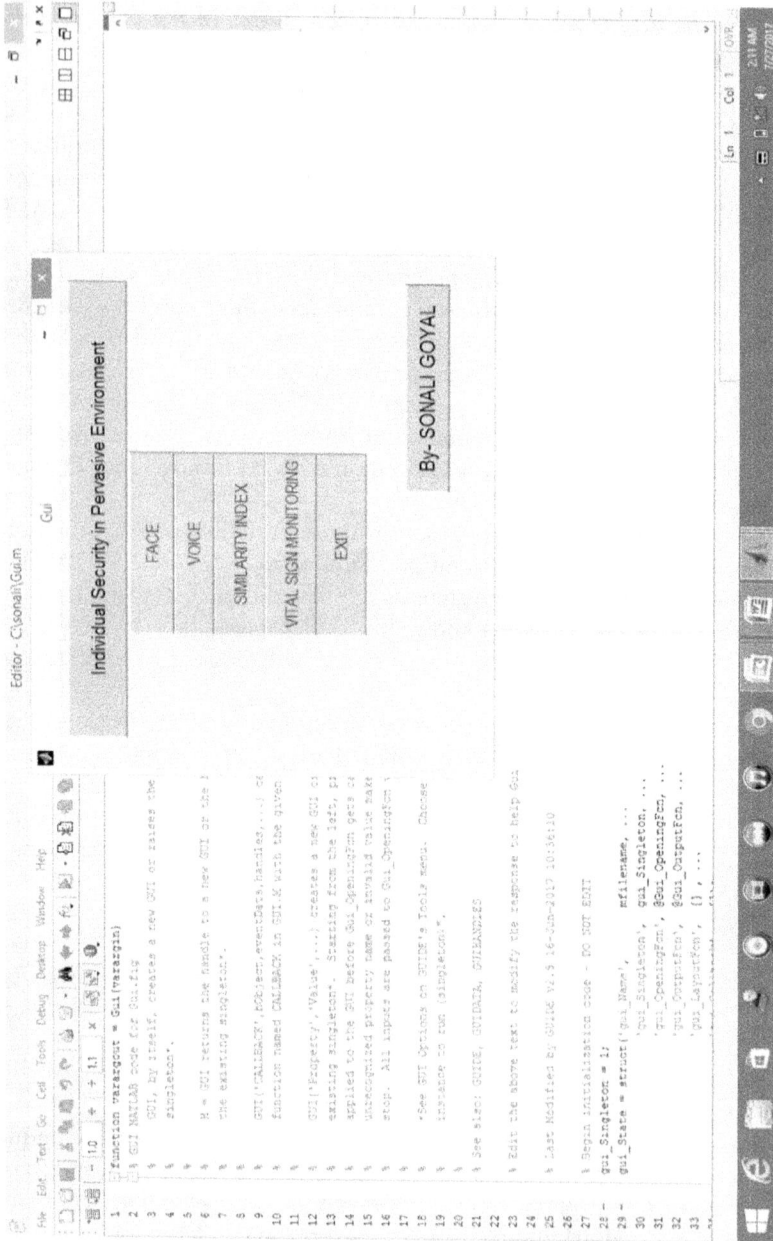

Figure 5.2 GUI of overall proposed model.

Table 5.1 Cases of overall proposed work

S. No	Face recognition	Voice recognition	Similarity index	Message
1	Yes	Yes	–	Category members
2	Yes	No	SI>=75%	Category members
3	No	Yes	SI>=75%	Category members
4	No	No	SI>=75%	Unknown

5.2 VITAL SIGN MONITORING

Next step is to monitor the health parameters of the individual living alone at home. In this step, the video in.mov format containing finger tip impression is browsed as shown in Figure 5.10. This video is recorded by iPhone 6s and then copied to laptop for further processing using MATLAB tool.

Figure 5.11 displays the sampling rate by processing all the frames of finger tip impression of an individual by using peak detection and smoothing function.

Figure 5.12 provides a graph for heart beat calculation, as it is shown in Figure 5.11 that the value for heart beat is equal to 58 bpm. It means heart beat is in normal range. In case, the value goes up or down from the specified values, then only the text message will be sent to all family members.

5.3 CONCLUSION AND FUTURE WORK

As the face recognition technique is hands free, it provides continuous authentication to the users. This technology captures the images through the camera and those features of the face are extracted which do not change with time. Similar to this, voice recognition plays a vital role in the authentication in which audio files are captured with the help of a microphone. The proposed model is implemented using face and voice recognition techniques to provide home based security to the individual living alone. The face recognition technique is used at the front door of the home in real time, even when no one is in the home and voice recognition technique is used to provide communication between the door and hardware device such as mobile or laptop. This system is very easy to handle and user friendly. This model incorporates a number of factors to provide safety and security in terms of a person's identification that who is at the door. Along with this implemented model, health monitoring aspect of owner of the home is also taken into consideration. A cost effective, easy to use model is developed which helps the users to continue to live in their comfortable environment at home. This model will allow healthcare personnel

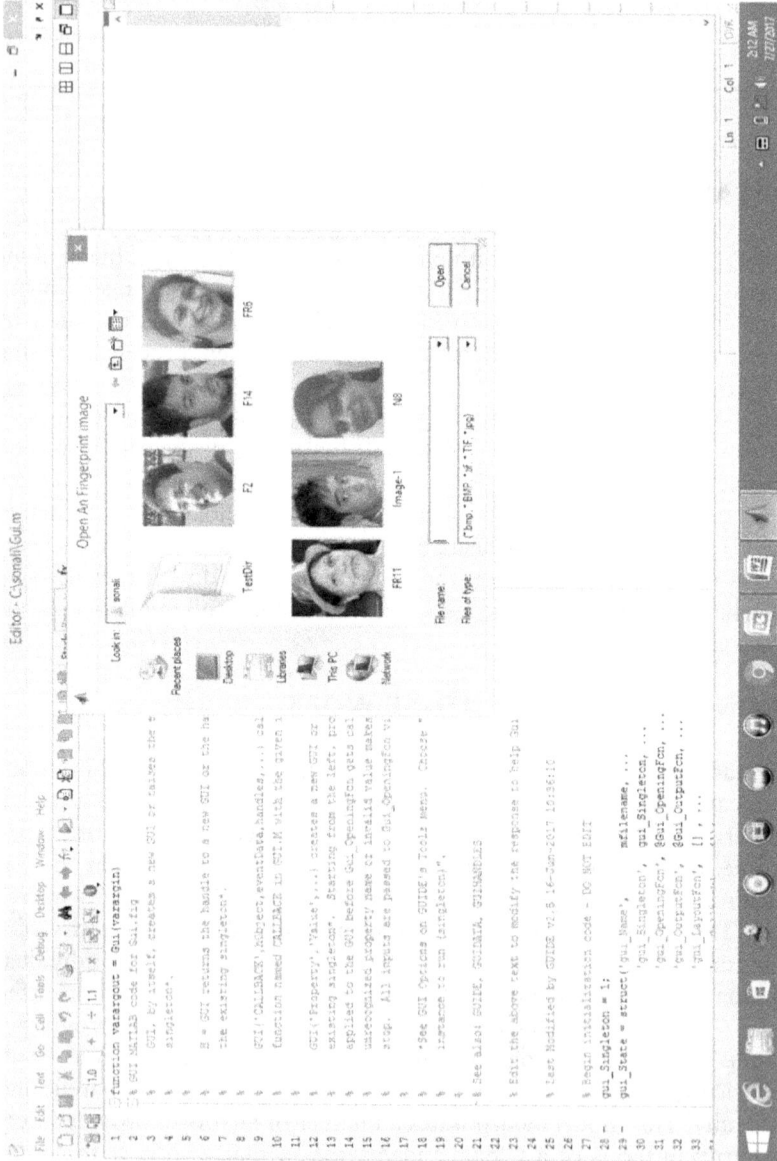

Figure 5.3 Face recognition process.

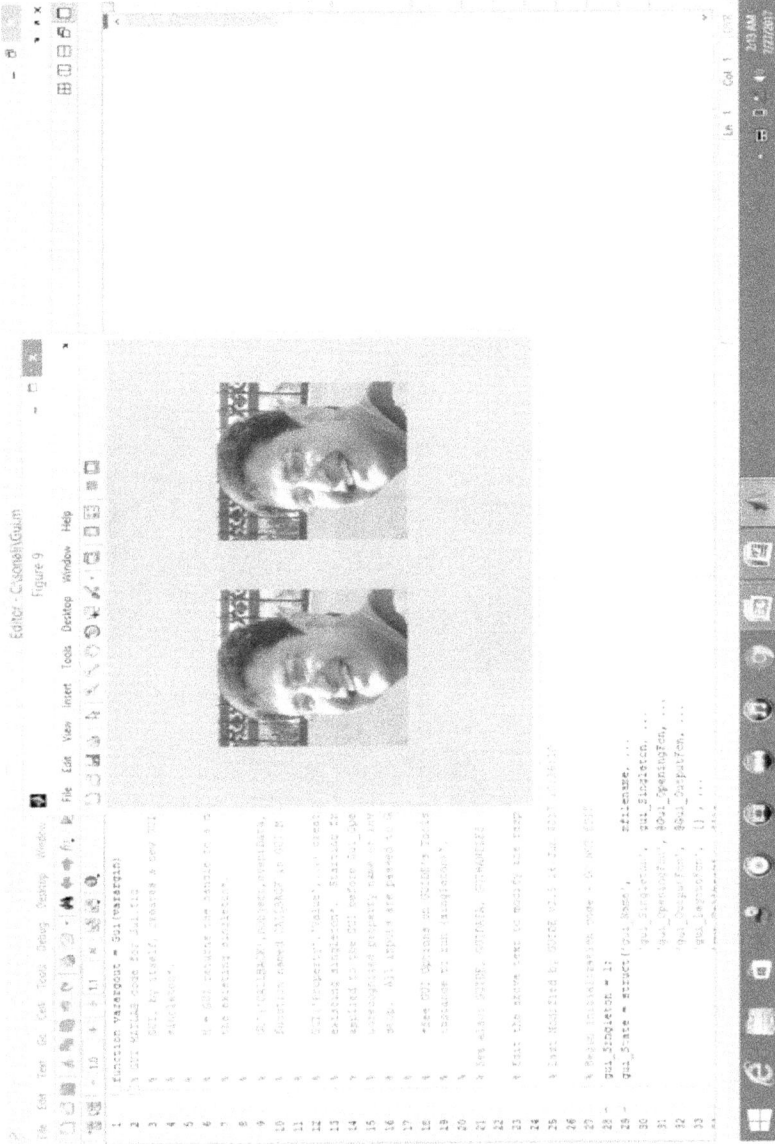

Figure 5.4 Comparison of an image with stored images in case of face recognition.

Figure 5.5 Output of face recognition with category display (family, friend, neighbor and unknown).

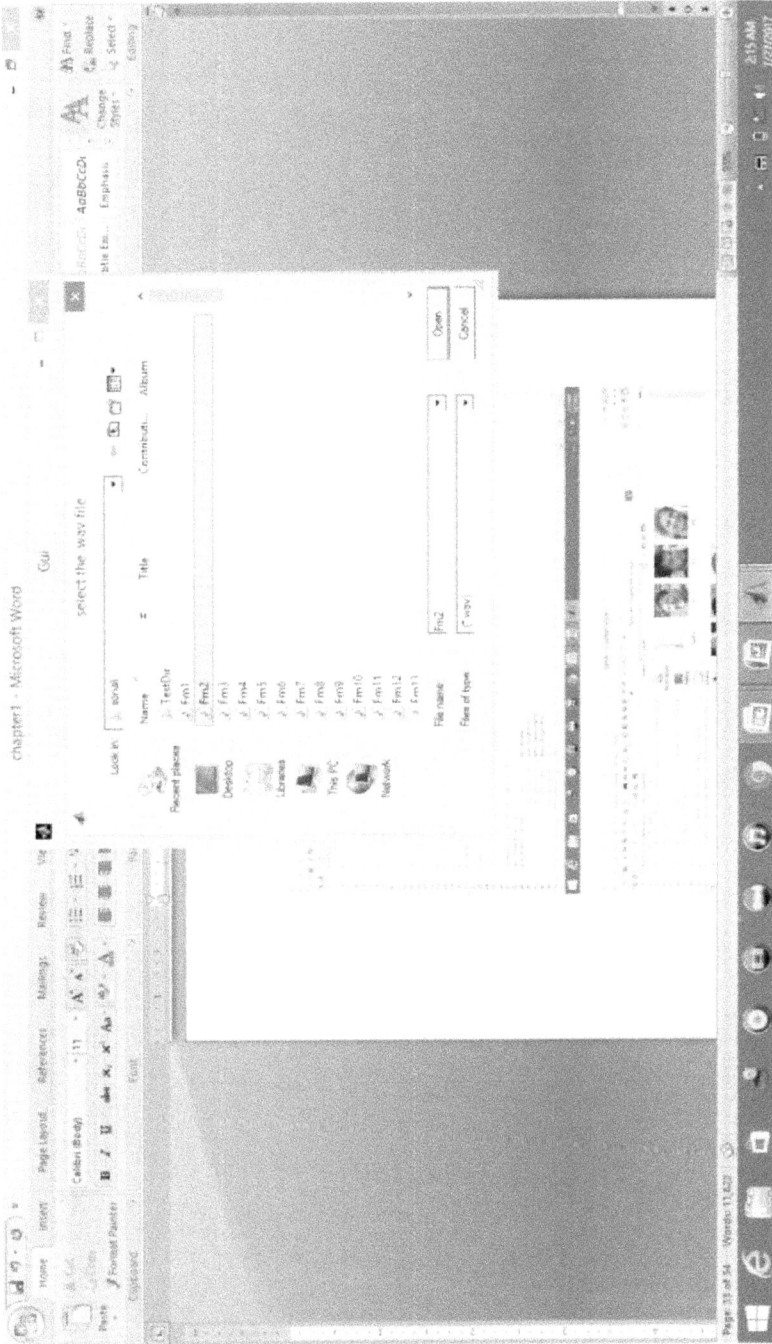

Figure 5.6 Voice recognition process.

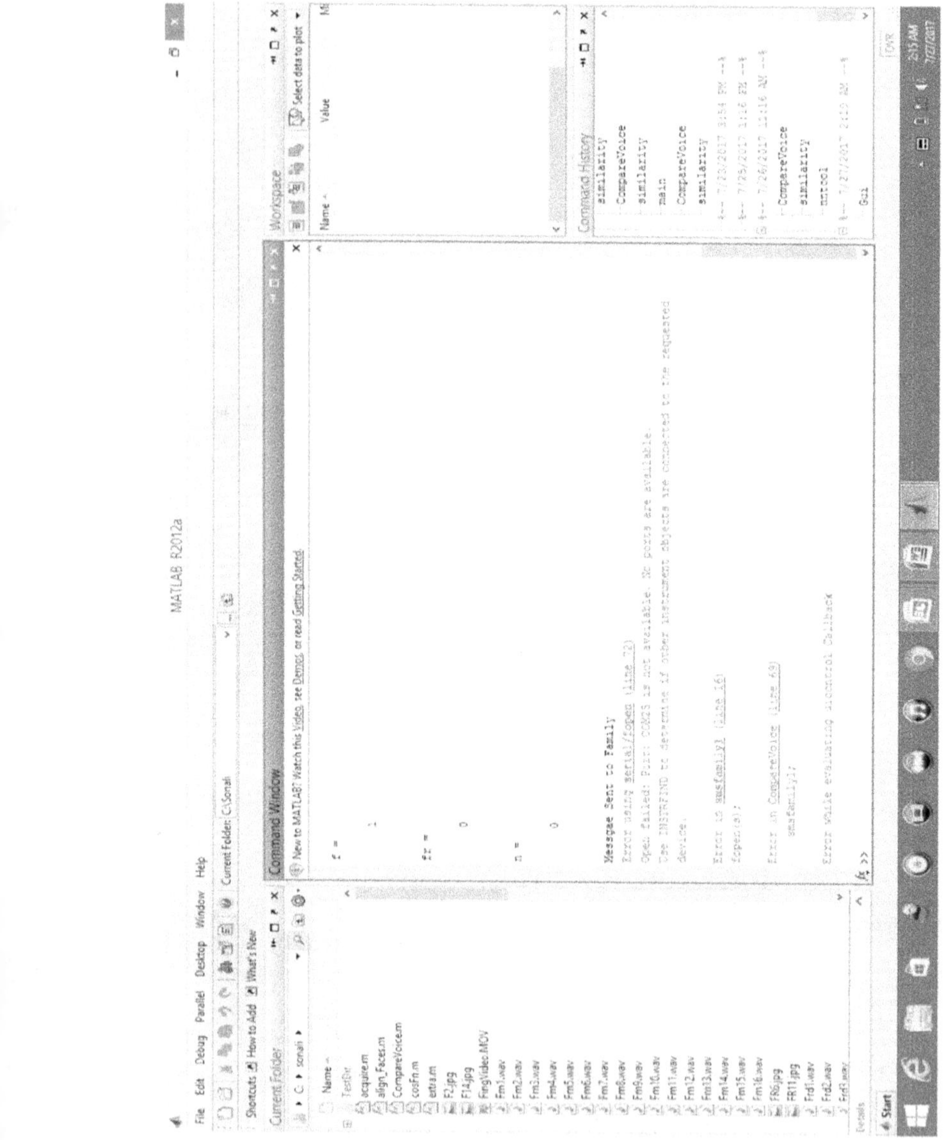

Figure 5.7 Output of voice recognition with category display (family, friend, neighbor and unknown).

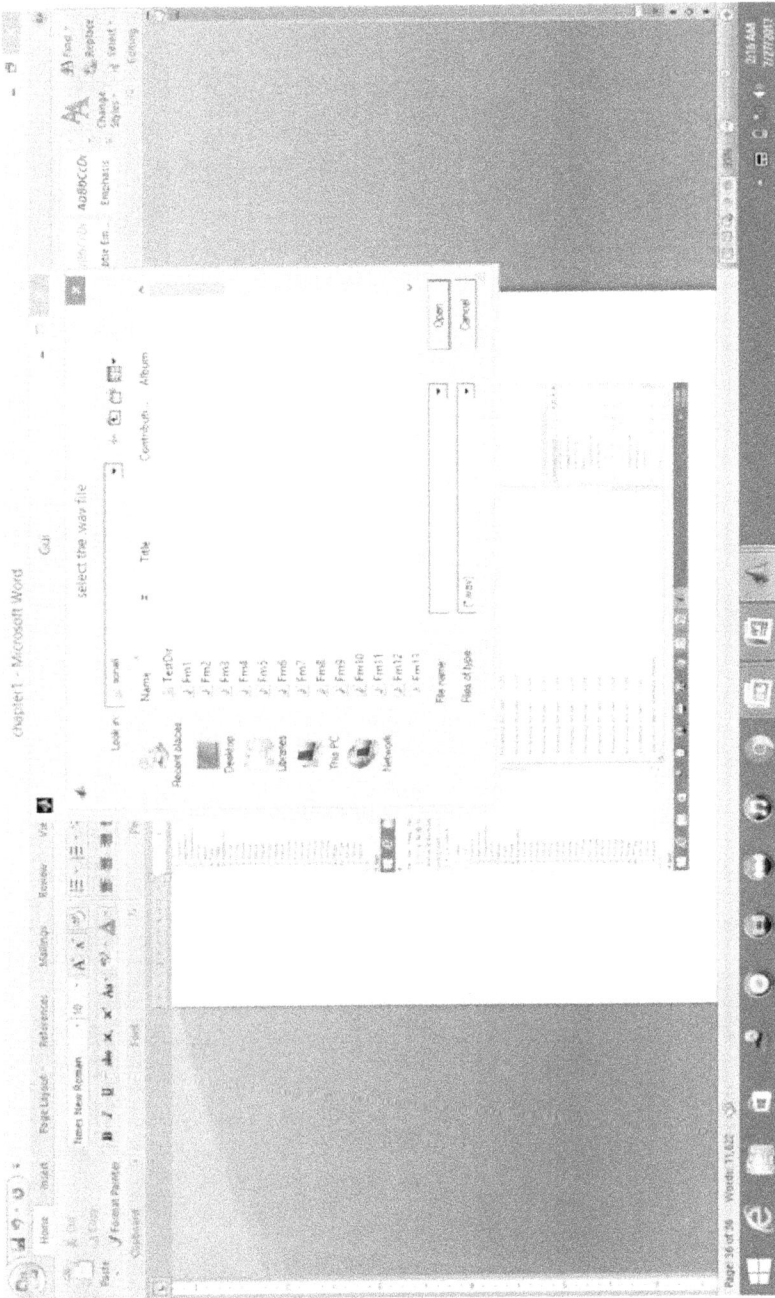

Figure 5.8 Similarity index calculation.

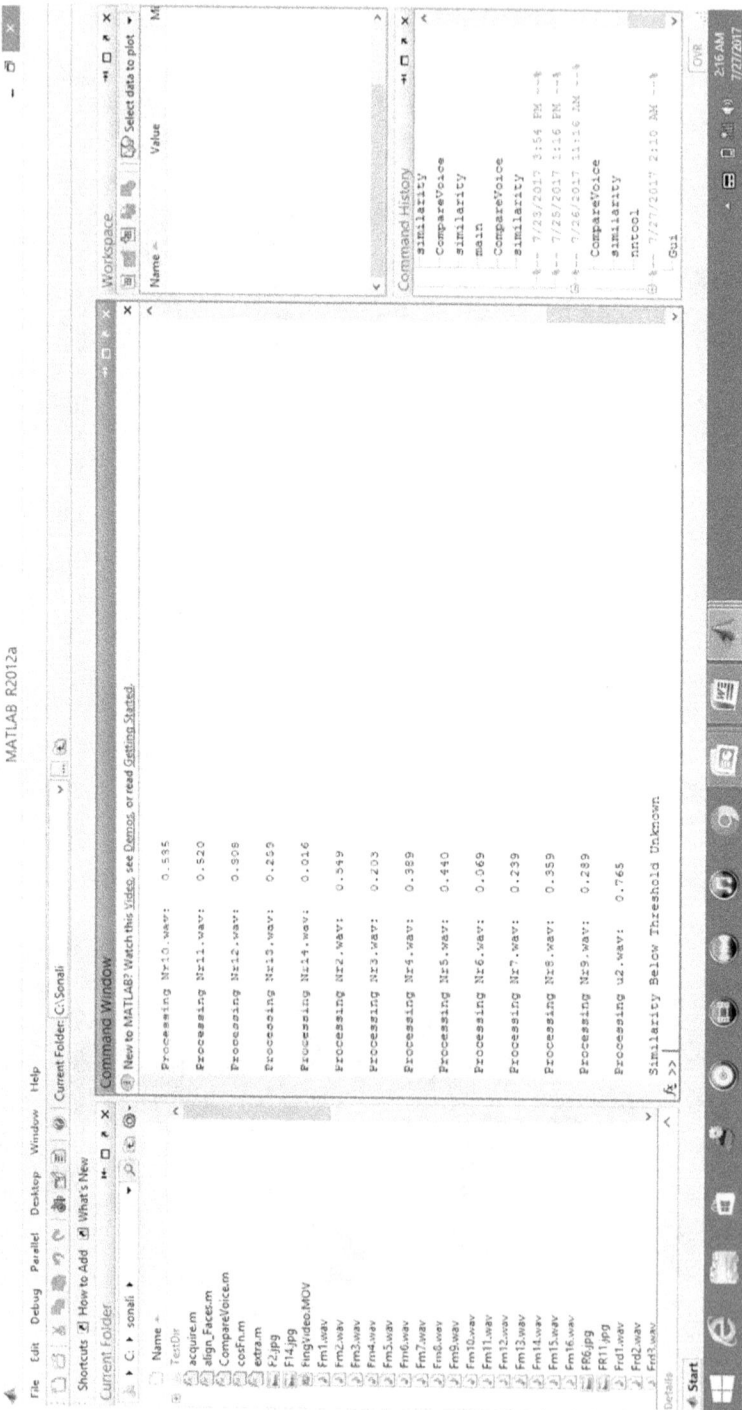

Figure 5.9 Output of similarity index in the form of percentage.

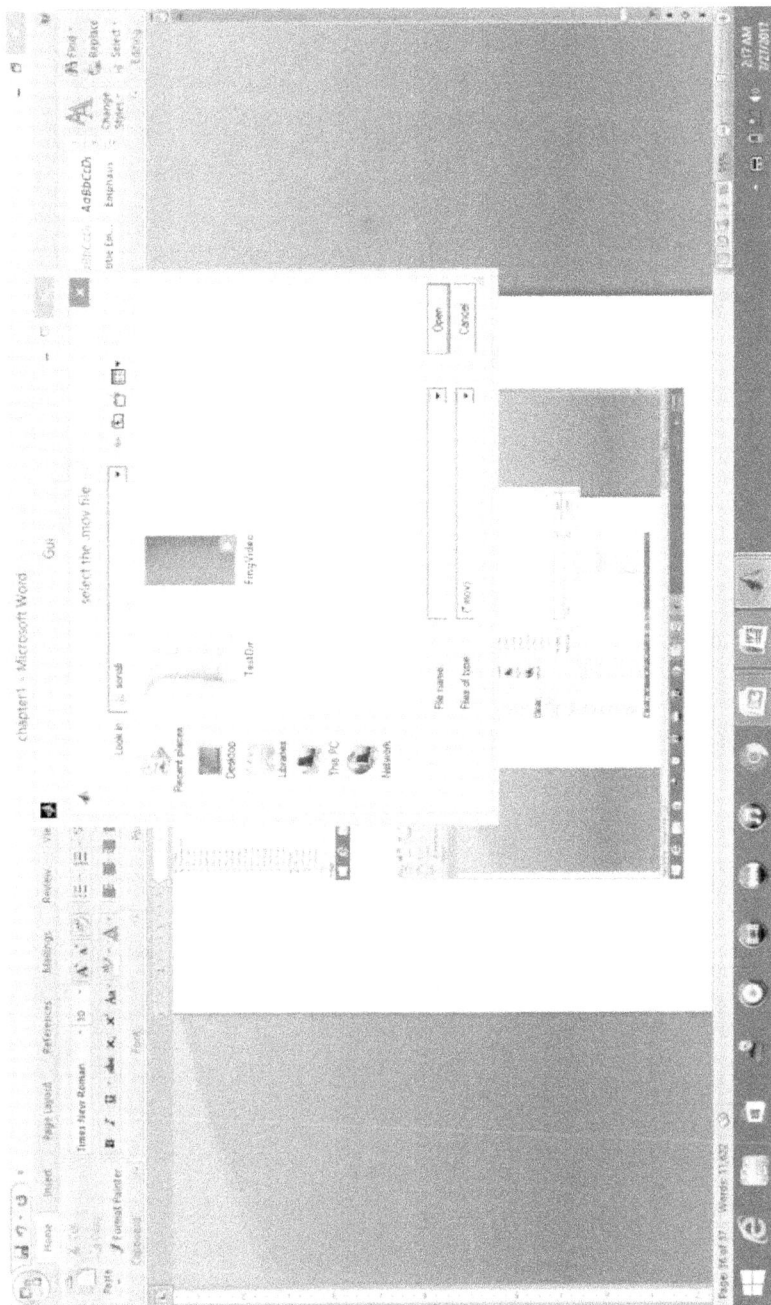

Figure 5.10 Video for monitoring of health parameters.

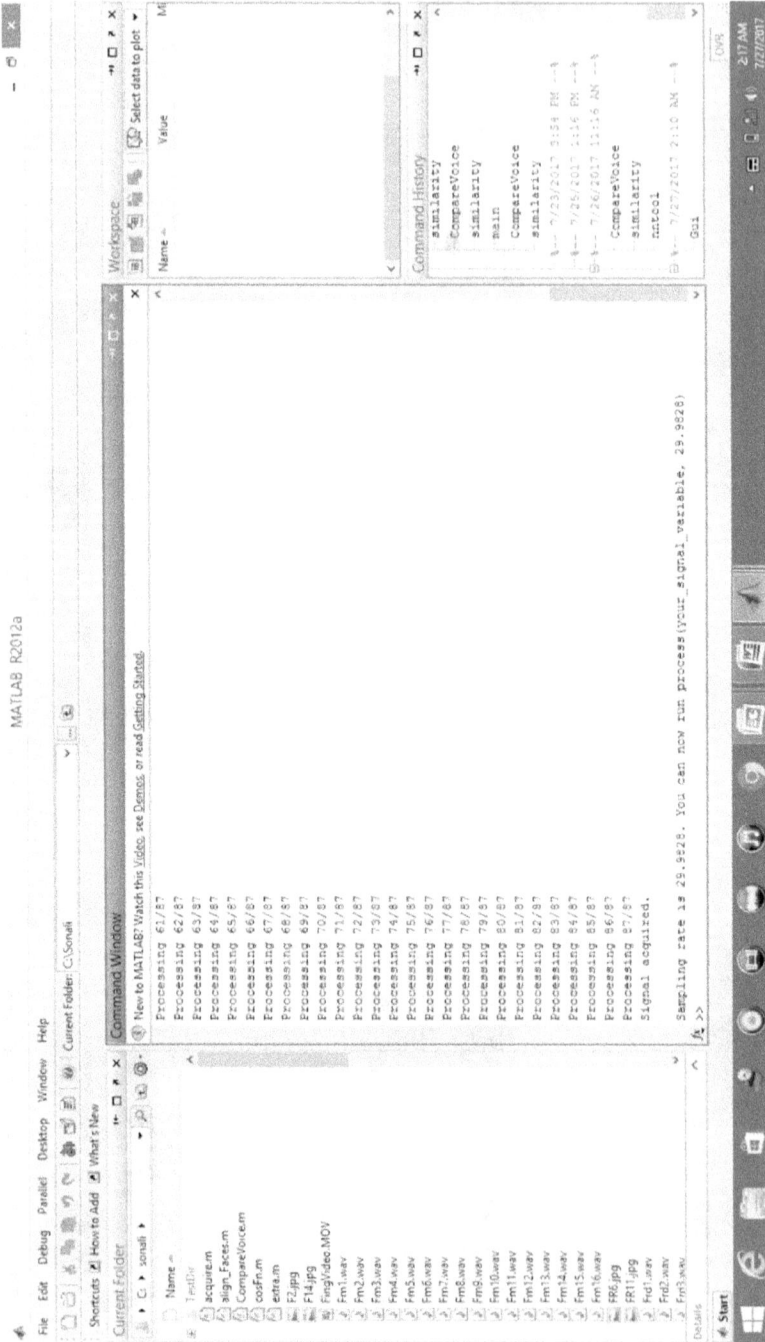

Figure 5.11 Output in form of sampling rate.

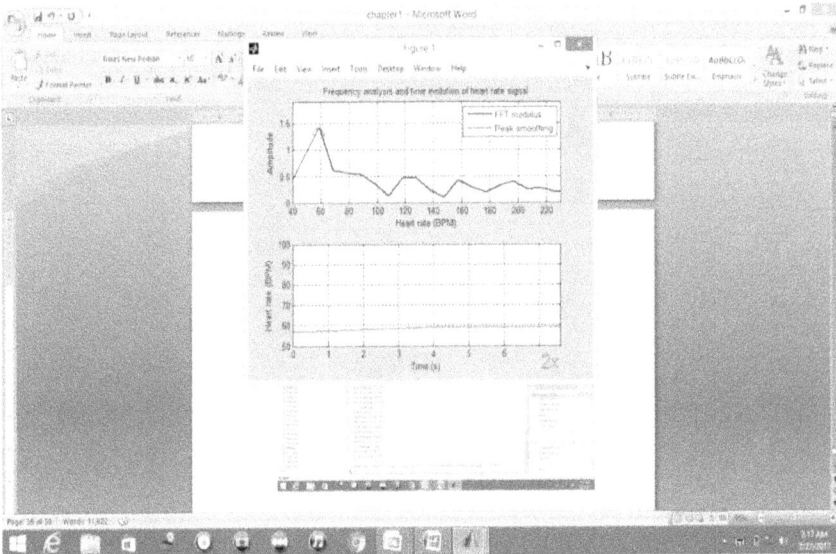

Figure 5.12 Graphs for heart beat monitoring.

to check important body signs of their patients in real time fix health conditions and provide feedback from distant facilities. In future, concept of soft computing can be used for the automatic face recognition system. Using the concept of fuzzy logic with neural networks, the performance of face recognition can be enhanced. Two or more classifiers can be combined to achieve better results. In case of voice recognition, the obtained results can be improved by fine tuning the system with larger training databases.

Chapter 6

Conclusion and future research

6.1 INTRODUCTION

As the old age states that the facial expressions are the one that completely describes about a person like his/her nature, emotional feeling, etc. That's why the face is termed as the complete index of the mind. So, the uniqueness of a particular person is recognized by considering the features of face of that particular person. To identify a person in a public area is not an easy task, and this is the reason that it is required to evolve efficient authentication techniques as hair style change, with spectacles, with hat, surgery of face can modify the natural look of face [79]. Sometimes, face recognition becomes a difficult step if the features of face change frequently. So, models must be developed to identify a person and recognize the face of that particular person for validation purpose. This model performs an important role in a number of applications such as security systems, banking and hotels. Face recognition system starts by locating a face in an image even in the changing situations such as change in pose, illumination and orientation. Face recognition methods consist of number of categories: a) knowledge-based: which is used for face localization. This method tries to automate user's knowledge of face recognition in the form of rules by using the concept of capturing the relationships between facial features, b) feature invariant: in which those structures of the facial features are captured that exist in case of change in pose, viewpoint, etc., and then these structures are used for face recognition process, c) template matching: in which complete face is used to take standard patterns and features of face are taken separately to compare an input image with number of images stored in a database, d) appearance-based: which also tries to acquire the varying features of all the stored images and then use these features of the stored set to recognize an input face, e) knowledge-based: number of rules are established that are based on the extraction process of information with help of research's knowledge [80].

Face recognition has taken proper form by using a wide range of areas. There are two types of face recognition methods that have been described

DOI: 10.1201/9781003120933-6

named as: feature based and holistic based. In feature based, features are extracted on the basis of properties and geometrical relationship is calculated on the basis of facial features including eye, chin, mouth and nose [81]. This approach is not suitable for practical applications. In holistic or appearance-based approach, full image representation is used, and it is based on high-dimensional vector representation. The face recognition system consists mainly of two steps: extraction of features with recognition of face. In extraction process, feature is extracted from face and in recognition; these features are matched with all the features stored in the database. Then, similarity score will be checked that is based on Euclidean distance and then used in taking the final decision. A variety of algorithms are used for extraction of features in a face recognition system named as: PCA, ICA, LDA, etc. but most commonly used method is a PCA (principle component analysis) with ICA (independent component analysis). They provides help in dimensionality reduction, they allow data compression while preserving the whole information; it converts all the images into vector form and store them in a single matrix [82], [83]. Sometimes, the performance of the system degrades when there is more variation in pose, illumination conditions, etc. So, this problem has been addressed in this proposed work. The hybrid PICA algorithm has been proposed in this research work to increase the recognition rate of the face recognition systems. Integrated PCA and ICA have been used for extraction of features and for classification: back propagation neural network is used. The performance for proposed algorithm in face recognition has been tested using 41 images in the training database. Hybrid PICA algorithm outperforms the previous methods. Voice recognition has also been performed that includes three main steps, i.e. processing of data, extraction of features and recognition of the voice [84]. Feature extraction is a process used to extract take out the features and convert them into feature vectors. Redundancy is removed from the features of a user. In every voice recognition system, there are two models: extraction and recognition phase. In the extraction phase, every model is trained by using features of voice and in recognition phase these features are matched with all the features placed in the database. By using this concept, similarity score will be calculated and this score is used to make the final decision. Voice recognition system can be classified in such a way that how many spoken words are recognized: isolated word needs each word is having silence on both the sides of the window. It operates on an individual word taken at a time with required pause between the words. It is the most simplified form of voice recognition as the endpoints are simpler to find. Connected words works in a similar way as isolated words, but it requires a minimum gap between the words separated to each other [85], [86]. Continuous speech is not separated by pauses; it works on speech in which all words are associated to each other. In this, a user speaks in a natural way, and it is difficult to recognize because it needs methods to define boundaries of words. There are a number of methods [87] that can be used for extraction of features such as MFCC, LPC, PLP and RASTA-PLP, but most commonly used methods are MFCC and PLP [88], [89]. They both carry speaker dependent as well as speaker independent information. But sometimes, the performance of voice recognition system degrades especially in noisy environments. So, this issue is addressed in the proposed work and the hybrid MFRASTA algorithm has been proposed with a way to increase the recognition rate of the voice recognition system. Algorithm comprises with four steps: first step is to collect data and apply pre-processing process in which audio files are recorded by number of methods in.wav format; feature extraction step extracts the features that provide specific information from the voice signals. Features have been extracted by using two voice recognition algorithms MFCC and RASTA-PLP, combine all the outputs and classify input voice that it is known or unknown and last is to send the message to define the category of each input voice. The hybrid MFRASTA gives improved results even in the utiquity of noise. The capability of proposed hybrid MFRASTA has been tested on a database consists of 41 audio files in.wav format. Another one is Similarity Index, which will run when one (face and voice) of the recognition technologies is not recognizing the user properly. It is text independent system in which a user is recognized on the basis of voice parameters such as pitch, tone and pronunciation, irrespective of the content of the words spoken by a user. Finally, health service providers [90] always face a challenge that how to provide comfortable health facilities for patients with low cost as the rapidly changing style of living in the present day world with the problem of a growing aging population requires a demanding need to update all facilities. So, the main objective is to establish a low cost and quickest way to provide health care to old age people. This research work presents all time available monitoring systems which will help in the monitoring of patients. There are a number of devices [91] that have been described for health monitoring such as phonocardiograph application, mercury sphygmomanometer and automatic digital sphygmomanometer, but the challenges with these devices have been addressed in the form of noise factor and bulkiness to be kept in a proper manner. Wearable sensors are the devices [92] that can be worn on the body as a part of accessories and can be used easily for health monitoring as every time it is not possible to go to hospital and in that case, it's quite hard to handle that situation. Numbers of methods are there to measure these parameters like PTT (pulse transit time) for BP estimation, but these methods require high complexity devices which are not convenient for users to wear. So, to address this issue, the main objective is to develop a low lost portable heart beat monitor device. A sensor-based fingertip application has been proposed in this proposed work by using an iPhone 6s camera sensor. With the help of this application, heart beat has been monitored in an accurate manner and results have been obtained by using MATLAB® tools for implementation and report has been sent to the family members in case of an emergency.

6.2 SUMMARY AND SCOPE

In Chapter 1, the meaning of pervasive computing, its challenges, uses of pervasive computing, issues involved in pervasive computing like a user is suddenly ill, abnormal heart rate, fall from sofa, abnormal BP, etc., and their effects have been discussed and explored [93]. This chapter further defines the concept of biometric technology, number of biometric techniques, basic types with their advantages and disadvantages in face recognition systems, its applications, basic structure of voice recognition system, its applications and two phases: speaker dependent and speaker independent system. Then, introduction to sensor-based technology, different types of sensors, different types of techniques used for health monitoring and uses of sensors for providing security to an individual is defined [94], [95]. Some basic concepts of neural network, its types, applications, number of uses and how it can be used for face and voice recognition is also stated. Finally, the meaning of MATLAB, all types of version and number of tools used for face and voice technology is explained. The issues and challenges of face recognition as well as voice recognition have been addressed to get the motivation for this thesis. The objective of this research work has been defined in this chapter.

It is clear that in the case of face recognition, different algorithms are defined for the extraction of features and number of neural network techniques is discussed for the recognition. Different types of face recognition methods that are used in face recognition tasks have been defined [96], [97]. Finally, dimensionality reduction technique has been discussed for providing high performance. In order to identify an unknown user, a large number of calculations are required to be done [98], [99]. Next is the voice recognition system, which describes all types of recognition techniques, the way to develop fast and effective algorithms and how to deal with noisy environment has been thoroughly explored and implemented [100], [101]. Different types of features of speech have been discussed [102], [103] with the use of different neural network techniques. In this sensor-based application, various previously used methods are discussed and a number of sensors have been defined for the health monitoring of users with a number of techniques used for sending messages to the users [104], [105]. A brief description of all the previously used methods for health monitoring as defined by various researchers in the conventional sensor-based system has also been presented. In this research work, the main focus has been on providing the security for an individual who is living alone at home by using multimodal biometric technology and sensor-based application.

In Chapter 2, process of face recognition has been defined by using hybrid PICA algorithm. For feature extraction, a combination of PCA + ICA has been described. Dimensionality reduction feature of PCA helps in removing the redundancy and then improves the performance of face recognition system [106]. The hybrid PICA algorithm helps in obtaining higher recognition rate as compared to individual PCA and individual ICA technique. Individual methods sometimes make wrong decisions in case of change in pose, illumination conditions, but the proposed algorithm provides high accuracy with all conditions. Back Propagation Neural Network is the most preferable network used for recognition with this hybrid algorithm. The complete implementation is done using image processing toolbox in MATLAB. Finally, a process of sending messages on the cell phone of family members using a 3G dongle with MATLAB tool is performed. In the proposed work, total 30 features have been extracted, and 41 images are taken from the training database in dimension of 512*512.

In Chapter 3, process of voice recognition is described. For feature extraction, MFCC and RASTA-PLP techniques have been used. RASTA-PLP helps in enhancing the performance of the recognition system even in noisy environment [107]. A hybrid algorithm named as MFRASTA has been proposed that shows the significant improvement in recognition rate. Previously, a number of combinations of voice recognition techniques have been used [108] for voice recognition system and a number of neural network techniques have been used with voice recognition techniques, but sometimes it makes the wrong decision and gives inaccurate results, but the proposed method outperforms them to achieve highest recognition rate with accurate results. This hybrid technique has been defined using text dependant as well as text independent method for speaking style, frequency and pitch [109], [110]. The database consists of about 41 audio files (in.wav format) to evaluate the proficiency of voice recognition system with different conditions. Experimental results clearly indicate that the performance of the hybrid work is greater than individual feature extraction techniques. The complete system is implemented in MATLAB tool with Graphical User Interface and finally this work is tested a number of times and found to be very reliable.

Chapter 4 describes a fingertip sensor-based application for heart beat monitoring. Heart rate is measured through a finger tip using optical sensors in a way to detect the blood flow from the finger and then the brightness of the acquired signals through the sensors is computed. In all previously used techniques, heart rate was considered by placing the thumb over arterial pulsation, timing and then count the pulses in 30 seconds [111]. Heart rate acquired is then multiplied by 2, this method seems to be very simple, but it provides errors in high rate. So, to overcome these problems, a new application has been proposed and implemented which measures heart beat through a fingertip. The basic objective of this research work is to implement a low-cost portable device that can be used anywhere, anytime and by any person. It also aims to improve the accuracy of current proposed application by identifying the drawbacks of existing techniques. This proposed application is implemented in MATLAB. It has a major advantage that non professional people can use this application at home for heartbeat measurement in a safe and easy way. This method provides accurate results even with a low-end video camera which is integrated in mobile devices.

In Chapter 5, final GUI is prepared which includes face recognition, voice recognition, similarity index and vital signs monitoring. All the steps are performed in a way to recognize a person at door and health monitoring of the individual at home is also performed. The complete steps are implemented in MATLAB tool. Face recognition is done by PICA algorithm and Voice recognition is done by MFRASTA algorithm. Vital signs monitoring is verified by fingertip application. Finally, messages will be sent to all category members belonging to different categories as per the requirement.

The outcomes of this work are compiled as-

- Recognition rate is improved by 99.5% due to the implementation of the hybrid PICA algorithm as compared to individual PCA and individual ICA in face recognition system.
- In case of voice recognition, text-dependant system is improved by 99% with implementation of hybrid MFRASTA techniques and text-dependant system is improved by 87% for known persons and 75% for unknown persons. Overall, this combination provides high recognition rate as compared to individual MFCC. The results indicate that in a noisy environment, this proposed MFRASTA is much better to use under robust conditions.
- The sensor-based fingertip application provides high accuracy for heart beat monitoring by using an iPhone 6s camera and in an emergency, send text messages to their family members. This proposed application is compared with a number of previously used techniques and finally it is observed that this method outperforms the previous methods.
- A real-time recognition system of face and voice is designed, and it gives better performance in terms of recognition rate. This complete system is implemented in MATLAB by creating a GUI and using a 3G dongle for sending the messages to the cell phone of the users.

6.3 FUTURE WORK

- When a specific category is displayed to which the person at door belongs after Face and Voice recognition, it will be displayed with the name of the person who appears at the door.
- The acceptable images are of 512*512 dimensions, they may be of any dimension and of any resolution.
- In case of health monitoring, some other parameters may be included such as stress rate, sweat rate and temperature.

- To ensure the exactness of the proposed heart rate monitoring device, number of experiments can be implemented by taking group of persons having different weights and with different age groups.
- To develop better algorithms and appropriate sensors to reduce acquisition rate failure.

References

1. Francesco Marcelloni, Daniele Puccinelli, Alessio Vecchio. Special Issue on Sensing and Mobility in Pervasive Computing. *Springer: Journal of Ambient Intelligence and Humanized Computing.* 2013;5(3):263–264.
2. Stan Kurkovsky. Pervasive Computing: Past, Present and Future. International Conference on Information and Communication Technology. 2008 Jan;1–7.
3. Kamal Sheel Mishra, Amit Kumar Tripathy. Some Issues, Challenges and Problems of Distributed Software System. *International Journal of Computer Science and Information Technologies.* 2014;5(4):4922–4925
4. M. Satyanarayan. Pervasive Computing Vision and Challenges. *IEEE Personal Communications.* 2001 Aug;8(4):10–17.
5. Gabor Kiss, David Sztaho, Klara Vicsi. Language Independent Automatic Speech Segmentation into Phoneme like Units on the Base of Acoustic Distinctive Features. IEEE International Conference on Cognitive Info Communications. 2013 Dec;10:579–582.
6. S. Lakshmikanth, K. R. Nataraj, K.R. Rekha. Noise Cancellation in Speech Signal Processing – A Review. *International Journal of Advanced Research in Computer and Communication Engineering.* 2014 Jan;3(1):5175–5186.
7. Nojun Kwak, Chong-Ho Choi, Jin Young Choi. Feature Extraction using ICA.*Lecture Notes in Computer Science.* Berlin: Springer. 2001;568–573.
8. Paolo Rosario Rao, Giulio Romano, Ardian Kita, Fernanda Irrera. Devices for the Real Time Detection of Specific Human Motion Disorders. *IEEE Sensors Journal.* 2016 Dec;16(23):8220–8224.
9. P. Dheerender, Ankita Chaturvedi, Sourav Mukhopadhyay. An Improved Biometric Based Remote User Authentication Scheme for Connected Healthcare. *International Journal of Adhoc and Ubiquitous Computing.* 2015;18(½):75–84.
10. Basma M. Mohammad El-Basioni, Sherine M. Abd El-Kader, Mahmoud Abdelmonim Fakhreldin. Smart Home Design Using Wireless Sensor Network and Bio-metric Technologies. *International Journal of Application or Innovation in Engineering and Management.* 2013;2(3):413–429.
11. Irvin Hussein Lopez Nava, Angelica Munoz Melendez. Wearable Inertial Sensors for Human Motion Analysis: A Review. *IEEE Sensors Journal.* 2015;16(22):7821–7834.

12. Andreas Komnios, Peter Barrie, Julian Newman. The Incidental Approach to Mobile Healthcare. IEEE International Conference on Pervasive Computing Technologies for Healthcare. IEEE. 2009.

13. Nawaf Hazim Barnouti, Sinan Sameer Mahmood, Muhammed Hazim Jaafer Al-Bamarni. Real Time Face Detection and Recognition Using PCA- BPNN and RBF. *Journal of Theoritical and Applied Information Technology.* 2016;91(1):28–34.

14. Manal Abdullah, Sahar, Majda Wazzan. Optimizing Face Recognition Using PCA. *International Journal of Artificial Intelligence and Applications.* 2012;3(2):23–31.

15. Gurpreet Kaur, Sukhvir Kaur, Amit Walia. Face Recognition Using PCA, Deep Face Method. *International Journal of Computer Science and Mobile Computing.* 2016 May;5(5):359–366.

16. Rahimeh Rouhi, Mehran Amiri, Behzad Irannejad. A Review on Feature Extraction Techniques in Face Recognition. *Signal and Image Processing: An International Journal.* 2012 Dec;3(6):01–14.

17. Hicham Mokhtari, Idir Belaidi, Said Alem. Performance Comparison of Face Recognition Algorithms Based on Face Image Retrieval. *Research Journal of Recent Sciences.* 2013 Dec;2:65–73.

18. Rahib H. Abiyev. Facial Feature Extraction Technique for Face Recognition. *Journal of Computer Science.* 2014;10(12):2360–2365.

19. Aseem Saxena, Amit Sinha, Shashank Chakrawati, Surabhi Charu. Speech Recognition Using MATLAB. *International Journal of Advances in Computer Science and Cloud Computing.* 2013 Nov;1(2):26–30.

20. Pratik K. Kurzekar, Ratnadeep R. Deshmukh, B. Vishal, Pukhraj P. Shrishrimal. A Comparative Study of Feature Extraction Techniques for Speech Recognition System. *International Journal of Innovative Research, Engineering and Technology.* 2014 Dec;3(12):18006–18016.

21. Beth Logan. Mel frequency Cepstral Coefficients for Music Modelling. International Symposium on Music Information Retrieval (ISMIR). Oct 2000.

22. Anchal Katyal, Amanpreet Kaur, Jasmeen Gill. Automatic Speech Recognition: A Review. *International Journal of Engineering and Advanced Technology.* 2014 Feb;3(3):71–74.

23. Mark Gales, Steve Young. The Application of Hidden Markov Models in Speech Recognition. *Foundations and Trends in Signal Processing.* 2007;1(3):195–304.

24. N. Easwari, P. Ponmuthuramalingam. A Comparative Study on Feature Extraction Technique For Isolated Word Speech Recognition. *International Journal of Engineering and Techniques.* 2015 Dec;1(6):108–116.

25. Narendra Kumar, Alok Aggarwal, Nidhi Gupta. Wearable Sensors for Remote Health Monitoring System. *International Journal of Engineering Trends and Technology.* 2012;3(1):37–42.

26. Marie Chan, Daniel Esteve, Christophe Escriba, Eric Campo. A Review of Smart Home: Present State and Future Challenges. *Elsevier: Computer Methods and Programs in Bio-medicines.* 2008 Mar;91(1):55–81.

27. Hoeyoung Kim, Chan-Hyun Youn, Dong-Hyun Kim, Hyewon Song, Eun Bo Shim. Adaptive Workflow Policy Based Management Scheme for Advanced Heart Disease Identification. IEEE International Symposium on Pervasive Systems, Algorithms and Networks. 2009;545–549.

28. Lara Srivastavas. Wireless Innovation for Smart Independent Living. Institute for Samfundsudvikling og Planlagning, Aalborg University. 2009.
29. Jayashri Bangali, Arvind Shaligram. Energy Efficient Smart Home based on Wireless Sensor Network Using LABVIEW. *American Journal of Engineering Research*. 2013;2(12):409–413.
30. Changu Suh. The Design and Implementation of Smart Sensor Based Home Networks. *IEEE Transactions*. 2008;54(3):1177–1184.
31. Rischan Mafrur, Priagung Khusumanegara, Gi Hyun Bang, Do Kyeong Lee, I Gde Dharma Nugraha, Deokjai Choi. Developing and Evaluating Mobile Sensing for Smart Home Control. *International Journal of Smart Home*. 2015;9(3):215–230.
32. Kulwinder Singh, Ashok Sardana, Kiranbir Kaur. Face Recognition by Plotting Different Facial Marks Using Neural Networks. *International Journal of Computer Science and Technology*. 2011 Jun;2(2):221–224.
33. Laurene V. Fausett. *Fundamentals of Neural Network: architectures, Algorithms and Applications*. United States: Prentice-Hall, 1994. 461.
34. Henry A. Rowley, Shumeet Baluja, Takeo Kanade. Neural Network Based Face Detection. *IEEE Computer Vision and Pattern Recognition*. 1998 Jan;20(1):23–38.
35. P. Melin, O. Castillo. Human Recognition Using Face, Fingerprint and Voice. *Hybrid Intelligent Systems for Pattern Recognition using Soft Computing*. Berlin: Springer, 2005;241–256.
36. Shamla Mantri, Kalpana Bapat. Neural Network Based Face Recognition Using MATLAB. *International Journal of Computer Science Engineering and Technology*. 2011 Feb;1(1):6–9.
37. Ryhan Ebad. Ubiquitous Computing: A Brief Review of Impacts and Issues. *MAGNT Research Report*. 2014;2(3):59–68.
38. R. Rameswari, S. Naveen Kumar, M. Abhishek Aananth, C. Deepak. Automated access control system using face recognition. *Materials Today*. May 2020.
39. Murat Taskiran, Nihan Kahraman, Cigdem Eroglu Erdem. Face Recognition: Past, Present and Future (a Review). 2020. Vol. 106. ISSN 1051-2004.
40. Sumanta Saha, Sharmishtha Bhattacharya. A Survey: Principal Component Analysis (PCA). International Conference on Innovative Trends in Science, Engineering and Management. 2017.
41. P. Shiva Prasad. Independent Component Analysis. Thesis. 2007.
42. Md. Shahjahan Kabir, Md. Solaiman Hossain, Dr. Rabiul Islam. Performance Analysis between PCA and ICA in Human Face Detection. *International Journal of Control, Automation, Communication and Systems*. 2016 Jan;1(1):25–34.
43. Suman Kumar Bhattacharya, Kumar Rahul. Face Recognition by Linear Discriminant Analysis. International Conference on Recent Advancements in Electrical, Electronics and Control Engineering. 2011.
44. Urvashi Bakshi, Rohit Singhal. A Survey on Face Detection Methods and Feature Extraction Techniques of Face Recognition. *International Journal of Emerging Trends and Technology in Computer Science*. 2014 June;3(3):233–237.

45. Hyeonjoon Moon, P. Jonathon Phillips. Computational and Performance Aspects of PCA Based Face Recognition Algorithms. *Perception.* 2001;30:303–321.

46. Nilind Sharma, Shiv Kumar Dubey. Face Recognition Analysis Using PCA, ICA and Neural Network. *International Journal of Digital Application and Contemporary Research.* 2014 Apr;2(9).

47. Qasim Alshebani, Prashan Premaratne, Peter Vial. A Hybrid Feature Extraction Technique for Face Recognition. International Conference on Information Technology and Science. 2014.

48. S. Satonkar Suhas, B. Kurhe Ajay, B. Prakash Khanale. Face Recognition Using PCA and LDA on Holistic Approach in Facial Images Database. *IOSR Journal of Engineering.* 2012 Dec;2(12):15–23.

49. Murad A.R., Anazida Z., Mohd. Aizaini Maarof. Principle Component Analysis Based Data Reduction Model for Wireless Sensor Networks. *Int. J. Adhoc and Ubiquitous Computing.* 2015;18(½):85–101.

50. Surya Kant Tyagi, Pritee Khanna. Face Recognition Using Discrete Cosine Transform and Nearest Neighbor Discriminant Analysis. *International Journal of Engineering and Technology.* 2012 Jun;4(3):311–315.

51. Sang Jean Lee, Sang bong Jung, Jang Woo Kwon, Seung Hong Hong. Face Detection and Recognition using PCA. Proceedings of IEEE Region 10 Conference. TENCON 99. pp. 84–87. Vol. 1. 1999.

52. Vivek Banerjee, Prashant Jain. Principle Component Analysis Based Face Recognition System Using Fuzzy C-means Clustering Classifier. *International Journal of Application and Contemporary Research.* 2013 Jan;1(6).

53. Jamal Hussain Shah, Muhammad Sharif, Mudassar Raza, Aisha Aseem. A Survey: Linear and Nonlinear PCA Based Face Recognition Techniques. *The International Arab Journal of Information Technology.* 2013 Nov;10(6): 536–545.

54. Zaid Abdi, Alkareem Alyasseri. Face Recognition Using Independent Component Analysis Algorithm. *International Journal of Computer Applications.* 2015 Sep;126(3):34–39.

55. Abhishek S. Parab, Amol Joglekar. Implementation of Home Security System Using GSM Module and Microcontroller. *International Journal of Computer Science and Information Technologies.* 2015;6:2950–2953.

56. Tarikul Islam, S.C. Mukhopadhyay, N.K. Suryadevara. Smart Sensor and Internet of Things: A Postgraduate Paper. *IEEE Sensors Journal.* 2015;17(3): 577–584.

57. Bruce A. Draper, Kyungim Baek, Marian Stewart Barlett, J. Ross Beveridge. Recognizing Faces with PCA and ICA. *Computer Vision and Image Understanding.* 2003;91(2003):115–137.

58. M. A. Hambali, R.G. Jimoh. Performance Evaluation of Principle Component Analysis and Independent Component Analysis Algorithms for Facial Recognition. *A Multidisciplinary Journal Publication of the Faculty of Science.* 2015;2:47–62

59. Marian Stewart Bartlett, Javier R. Movellan, Terrence J. Sejnowski. Face Recognition by Independent Component Analysis. *IEEE Transactions on Neural Networks.* 2002 Nov;13(6):1450–1464.

60. Anil Kumar Vuppala, K.V. Mounika, Hari Krishna Vyadana. Significance of Speech Enhancement and Sonorant Regions of Speech for Robust Language Identification. IEEE International Conference on Signal Processing, Communication and Energy Systems. 2015.

61. Rohita P. Patil, Mohammdjaved R. Mulla. A Review: Design and Implementation of Image Acquisition and Voice Based Security System. *International Journal of Advanced Research in Electrical, Electronics and Instrumentation Engineering.* 2015 Mar;4(3).

62. Suma Swamy, K.V. Ramakrishnan. An Efficient Speech Recognition System. *Computer Science and Engineering: An international Journal.* 2013;3(4):21–27.

63. Varsha Singh, Vinay Kumar, Neeta Tripathy. A Comparative Study of Feature Extraction Techniques for Language Identification. *International Journal of Engineering Research and General Science.* 2004;2(3):286–291.

64. Harshita Gupta, Divya Gupta. LPC and LPCC method of feature extraction in speech recognition system. IEEE International Conference- Cloud System & Big Data Engineering. 2016

65. Aseem Saxena, Amit Kumar Sinha, Shashank Chakarawati, Surabhi Charu. Speech Recognition Using MATLAB. *International Journal of Advances in Computer Science and Cloud Computing.* 2013 Nov;1(2):26–30.

66. Smita B Magre, Ratnadeep R. Deshmukh. A Comparative study on Feature Extraction Techniques in Speech Recognition. International Conference on Recent Advances in Statistics and Their Applications. 2013.

67. Shirikant Upadhyay, Sudhir Kumar Sharma, Aditi Upadhyay. Analysis of Different Classifier Using Feature Extraction in Speaker Identification and Verification under Adverse Acoustic Condition for Different Scenario. *International Journal of Innovations in Engineering and Technology.* 2016 Apr;6(4):425–434.

68. Thea Franke, Catherine Tong, Maureen C. Ashe. The Secrets of Highly Active Older Adults. *Journal of Aging Studies.* 2013;27(4): 398–409.

69. Lucian Pestritu, Alexandra Todiruta, Maria Goga, Nicolae Goga. Method for Measuring the Heart Rate Through Fingertip Using a Low end Video Camera and its Applications in Self Care. *IEEE.* 2013:1–4.

70. T. Pursche, J. Krejewski. Video Based Heart Rate Measurement from Human Faces. IEEE International Conference on Consumer Electronics. 2012;544–546.

71. Julia Jansson. Developing a Method of Measuring Blood Pressure using a Video Recording. *Research Academy for Young Scientists.* 2016. Thesis.

72. Basma M. Mohammad El-Basioni, Sherine M. Abd El-Kader, Mahmoud Abdelmonim Fakhreldin. Smart Home Design Using Wireless Sensor Network and Bio-metric Technologies. *International Journal of Application or Innovation in Engineering and Management.* 2013;2(3):413–429.

73. Daniel H. Wilson. Assistive Intelligent Environments for Automatic Health Monitoring. 2005 Sep. Thesis.

74. Trupti M.K., Dr. Prashant R.M. Hybrid Approach to Face Recognition System using Principle Component Analysis and Independent Component Analysis with Score Based Fusion Process. *Computer Vision and Pattern Recognition.* 2014.

75. Milos Oravec, Jasmila Pavlovicova. Face Recognition Methods Based on Feed Forward Neural Network, PCA and SOM. *Radio Engineering.* 2007 Apr;16(1):51–57.

76. Vivek Banerjee, Prashant Jain. PCA Based Face Recognition System Using Fuzzy C-means Clustering Classifier. *International Journal of Digital Application and Contemporary Research.* 2013 Jan;126(3):34–38.

77. Kuldeep Kumar, Ankita Jain. A Hindi Speech Recognition System for Connected Words using HTK. *International Journal of Computational Systems Engineering.* 2012;1(1):25–32.

78. Tara S. Sainath, Brian Kingsbury, Hagen Soltau, Bhuvana Ramabhadran. Optimization Techniques to Improve Training Speed of Deep Neural Networks Long Large Speech Tasks. *IEEE Transactions on Audio, Speech and Language Processing.* 2013 Nov;21(11):2267–2276.

79. Tomas Matlovic, Peter Gaspar, Robert Moro, Jakub Simko, Maria Bielikova. Emotions Detection using Facial Expressions Recognition and EEG. *IEEE:*19–25.

80. X. Zhou, B. Bhanu. Feature Fusion of Side Face and Gait for Video Based Human Identification. *Pattern Recognition.* 2008;41:778–795.

81. G. Sasikumar, B.K. Tripathy. Design and Implementation of Face Recognition System in MATLAB Using the Features of Lips. *International Journal Intelligent Systems and Applications.* 2012;8:30–36.

82. Sukhvinder Singh, Meenakshi Sharma, N. Suresh Rao. Accurate Face Recognition Using PCA and LDA. International Conference on Emerging Trends in Computer and Image Processing. 2011;10(3):103–112.

83. K. Kim. Intelligent Control System by Using Passport Recognition and Face Verification. *Springer: International Symposium on Neural Networks.* 2005;3497:147–156.

84. Shweta Sinha, Aruna Jain. Spectral and Prosodic features based Speech Pattern Classification. *International Journal of Applied Pattern Recognition.* 2015;2(1):96–109.

85. Yulan Liu, Pengyuan Zhang, Thomas Hain. Using Neural Network Front Ends on Far Field Multiple Microphones Based Speech Recognition. IEEE International Conference on Acoustic Speech and Signal Processing. 2014 Oct; 5542–5546.

86. Tu Yanhui, Du Jun, Xu Yong, Dai Lirong, Lee Chin-Hui. Deep Neural Network Based Speech Separation for Robust Speech Recognition. *IEEE.* 2014;532–536.

87. Darryl Stewart, Rowan Seymour, Adrian Pass, Ji Ming. Robust Audio Visual Speech Recognition under Noisy Audio Video Conditions. *IEEE Transactions on Cyber netrics.* Feb 2014;44(2):175–184.

88. S. Lakshmikanth, K.R. Natraj, K.R. Rekha. Noise Cancellation in Speech Signal Processing- A Review. *International Journal of Advanced Research in Computer and Communication Engineering.* 2014 Jan;3(1):5175–5186.

89. B. Sudhakar, R. Bens Raj. Automatic Speech Segmentation to Improve Speech Synthesis, Performance. International Conference on Circuits, Power and Computing Technologies. 2013;835–839.

90. Lei Clifton, David A. Clifton, Marco A.F. Pimentel, Peter J. Watkinson, Lionel Tarassenko. Predictive Monitoring of Mobile Patients by Combining Clinical Observations with Data from Wearable Sensors. *IEEE Journal of Biomedical and Health Informatics.* 2014 May;18(3):722–731.

91. S. Gayathri, N. Rajkumar, V. Vinoth Kumar. Human Health Monitoring System Using Wearable Sensors. *International Research Journal of Engineering and Technology.* 2015;2(8):122–126.
92. Narendra Kumar, Alok Aggarwal, Nidhi Gupta. Wearable Sensors for Remote Health Monitoring System. *International Journal of Engineering Trends and Technology.* 2012;3(1):37–42.
93. U. Cortes, C. Borrue, A.B. Martinez. Assistive Technologies for the New Generation of Senior Citizens: The SHARE-It Approach. *Int. J. Computers in Healthcare.* 2010;1(1):35–61.
94. S. Rihana. Wearable Fall Detection System. Middle East Conference on Biomedical Engineering. 2016.
95. Elmusod Talipov, Yohan Chon, Hojung Cha. Content Sharing over Smartphone Based Delay Tolerant Networks. *IEEE Transaction on Mobile Computing.* 2013 Mar;12(3):581–595.
96. Tom Martin, Kahyun Kim, Jason Forsyth, Lisa McNair, Eloise Coupey, Ed Dorsa. Discipline Based Instruction to Promote Interdisciplinary Design of Wearable and Pervasive Computing Products. *Personal and Ubiquitous Computing.* 2011;17(3):465–478.
97. Rizoan Toufiq. *Face Recognition using Multiple Classifier Fusion.* LAP Lambert Academic Publishing Germany, 2012.
98. Pawel Krotewicz, Wojciech Sankowski, Piotr Stefan Nowak. Face Recognition based on 2D and 3D Data Fusion. *Int. J. Biometrics.* 2015;7(1):69–81.
99. Ajay Jaiswal, Nitin Kumar, R.K. Aggarwal. Analysis and Evaluation of Regression Based Methods for Facial Pose Classification. *International Journal of Applied Pattern Recognition.* 2015;2(1):24–45.
100. K-C. Kwak, W. Pedrycz. Face Recognition Using an Enhanced ICA Approach. *IEEE Transactions on Neural Networks.* 2007;18:530–541.
101. Ishwar S. Jadhav, V. T. Gaikwad, Gajanan U. Patil. Human Identification Using Face and Voice Recognition. *International Journal of Computer Science and Information Technologies.* 2011;2:1248–1252.
102. Dona Varghese, Domonic Mathew. Phoneme Classification Using Reservoirs with MFCC and RASTA-PLP Features. IEEE International Conference on Computer Communication and Informatics. 2016 Jan;1–6.
103. A. Milton, S. Sharmy Roy, S. Tamil Selvi. SVM Scheme for Speech Emotion Recognition Using MFCC Feature. *International Journal of Computer Applications.* 2013 May;69(9):34–39.
104. Shi Huang, Yu-Ren Luo Chen. Speaker Verification Using MFCC and Support Vector Machine. *International Multiconference of Engineers and Computer Scientists.* 2009;1:532–535.
105. Suryo Wijayo, Thiang. Speech Recognition Using Linear Predictive Coding and Artificial Neural Network for Controlling of Mobile Robot. International Conference of Information and Electronics Engineering. 2011;6(2011):179–183.
106. J.N.K. Liu, M. Wang, B. Feng. iBot Guard: An Internet Based Intelligent Robot Security System Using Invariant Face Recognition Against Intruder. *IEEE Transactions on Systems Man and Cybernetics Part- Applications and Reviews.* 2005;35:97–105.

107. Tara N. Sainath, Brian Kingsbury, Hagen Soltau, Bhuvana Ramabhadran. Optimization Techniques to Improve Training Speed of Deep Neural Networks for Large Speech Tasks. *IEEE Transactions on Audio, Speech and Language Processing*. 2013 Nov;21(11):2267–2276.

108. Samir Akrouf, Belayadi Yahia, Mostefai Messaoud, Youssef Chahir. A Multimodal Recognition System Using Face and Speech. *International Journal of Computer Science Issues*. 2011;8(1):1694–0814.

109. Abdelmajid Hassan Mansour, Gafar Zen Alabdeen Salh, Hozayfa Hayder Zeen Alabdeen. Voice Recognition Using Back Propagation Algorithm in Neural Networks. *International Journal of Computer Trends and Technology*. 2015 May;23(3):132–139.

110. Mohammed Hayyan Alsibai, Hoon Min Siang. A Smart Driver Monitoring System Using Android Application and Embedded System. IEEE International Conference on Control System, Computing and Engineering. 2015 Nov;242–246.

111. Prosanta Gope, Tzonelih Hwang. BSN Care: A Secure IoT-Based Modern HealthCare System Using Body Sensor Network. *IEEE Sensors Journal*. 2015;16(5):1368–1376.

112. Vahid Rafe, Masoumeh Hajvali. Designing an Architectural Style for Pervasive Healthcare Systems. *Journal of Medical Systems*. 2012;38(19):1–10.

Index

accelerometer, 38, 133
accessibility, 92
accuracy, 94
active tags, 39
adaptive training, 44
align faces, 85
ambient computing, 6
appearance-based, 157
artificial intelligence approach, 61
augmented reality, 6
automatic digital
        sphygmomanometer, 131

band pass filtering, 135
behavioural biometrics, 25
bias function, 63
biometrics, 14, 15, 16, 34
blood pressure, 126
bluetooth, 1, 2
brightness signal computation, 135

CDMA, 1, 2
centering, 67, 74
Cepstrum, 94
cloud computing, 4
cluster, 65
computation method, 44
context awareness, 9
continuous speech, 94
convolution, 56
cooperation unwillingness, 34
covariance matrix, 74

data acquisition, 59
DCT (Discrete Cosine Transform),
      23, 70
decoding, 28

desktop type computing, 1
device heterogeneity, 10, 11
distinctness, 92
DNA, 15, 32
DTW (dynamic time warping), 103
dynamic signature verification, 25

Eigen faces, 20, 63
electrocardiographs, 132
emergency alarm, 40
enrolment session, 26
Euclidean distance, 60
exactingness, 34

face recognition, 17, 19, 20, 59, 62, 141
facial thermograms, 30
Fast Fourier Transform, 136
FDA (Fisher Discriminant Analysis), 69
feature extraction, 18, 27, 92
feature matching, 92
Feed Forward Neural Network, 44
FFT (Fast Fourier Transform), 98
filters, 56
fingerprint, 17
fingertip, 134, 141
Fisher face, 23
Fourier transform, 56

Gabor Wavelet, 23, 71
Google voice, 92
GPS, 124
GSM, 1, 2

hand geometry, 24
healthcare, 7, 37
histogram, 55
HMM (Hidden Markov Model), 61

For Product Safety Concerns and Information please contact our EU
representative GPSR@taylorandfrancis.com
Taylor & Francis Verlag GmbH, Kaufingerstraße 24, 80331 München, Germany

www.ingramcontent.com/pod-product-compliance
Lightning Source LLC
Chambersburg PA
CBHW070721220326
41598CB00024BA/3253